HISTORIC CUMBERLAND COUNTY SOUTH

Land of Promise

ROGER DAVID BROWN

NIMBUS
PUBLISHING

Copyright © Roger David Brown, 2002

All rights reserved. No part of this book may be reproduced, stored in a retrieval system or transmitted in any form or by any means without the prior written permission from the publisher, or, in the case of photocopying or other reprographic copying, permission from CANCOPY (Canadian Copyright Licensing Agency), 1 Yonge Street, Suite 1900, Toronto, Ontario M5E 1E5.

Nimbus Publishing Limited
PO Box 9166
Halifax, NS   B3K 5M8
(902) 455-4286

Printed and bound in Canada

Designer: Terri Strickland

Front cover: The Maple Sugar Camp, East Mapleton, 1909

Title page: Parrsboro Blockhouse above Partridge Island Settlements, 1836

National Library of Canada Cataloguing in Publication Data

    Brown, Roger David
    Historic Cumberland County South : land of promise
    (Images of our past)
    ISBN 1-55109-418-5

1. Coal mines and mining–Nova Scotia–Cumberland (County)–History. 2. Coal trade–Nova Scotia–Cumberland (County)–History.
I. Title. II. Series.

TN806.C22N6 2002   338.2'724'0971611   C2002-903516-3

We acknowledge the financial support of the Government of Canada through the Book Publishing Industry Development Program (BPIDP) and the Canada Council for our publishing activities.

This book is dedicated in memory of Arthur Lee Dickinson of Parrsboro, Carl and Gladys Brown of Southampton, Nevin Burke of Amherst, and the students of Parrsboro and Advocate regional high schools.

October 29, 2002

To Anne,

Roger Brown

Contents

Prologue
    The Natural Word, The Human World vi

Chapter 1
    The Coal Fields by the Sea
    Joggins, River Hebert, Maccan, Chignecto 1

Chapter 2
    The Elysian Fields
    Minudie 37

Chapter 3
    The Fertile Inland Valleys
    Southampton and Newville 53

Chapter 4
    Where Dinosaurs Walked
    Parrsboro 75

Chapter 5
    God's Jewel Box
    Advocate Harbour to the Parrsborough Shore 105

Chapter 6
    The Town That Would Not Die
    Springhill 127

Epilogue
    There is a Future Here 147

Appendix A
    A List of Deaths in the Mines by the Sea 149

Appendix B
    Population Figures 153

Appendix C
    Springhill Mine Deaths 155

Acknowledgements 157

Image Sources 159

Prologue

THE NATURAL WORLD

Geology, geography, and history have woven the pattern of southern Cumberland county, which can be divided into three main geographical areas.

The first geographical division is the coal-bearing Carboniferous strata, laid down circa 350 million B.C.—50 million years before the start of the Mesozoic dinosaur era. Because Nova Scotia was then in an equatorial location, its great swamps were filled with vegetation. Periodically these swamps were buried by silt from the rushing rivers and from changes in ocean levels. Slowly, the buried plant life turned into coal.

The county's mines reveal many fossils. R. Morrow wrote in the nineteenth century:

> You can find the petrified remains of trees, ferns and various plants, often in profusion embedded in shale forming the immediate covering of the coal that has been removed from the mine...they assume various forms, some standing upright as if they grew where they are now, others leaning in posture, while a few are lying on their sides.

The cliffs around Joggins are famous for fossils, which become exposed in the process of erosion. In 1842, The Geological Survey of Canada was formed. One year later, the Joggins cliffs were the subject of a field project led by the survey's first director, Sir William Logan. (Canada's highest peak, Mt. Logan, Yukon, bears his name.) In 1851, Sir William Dawson and Sir Charles Lyell found fossils on the Joggins beaches and cliffs. In his writings, Dawson popularized the area. In this carboniferous zone, three towns would grow up: Springhill, Joggins, and River Hebert.

Between this region and the Parrsborough Shore lies the second geographical area: the eroded remains of the Appalachian Mountains, whose rounded tops form the county divide for water drainage. The mountains were created some 270 million years ago by shifting of the Earth's tectonic plates. Cutting through this range—known locally as the Cobequids—is the Parrsboro Gap, where long ago a giant river churned its way north from Hants County. Though today it is filled with rubble from "recent" glacial times (which ended about 10,000 B.C.), this pass provided the transportation route for road and rail.

The third major area is the Parrsborough Shore. Over 200 million years ago, the ancient supercontinent we now call Pangaea split apart, forming a great dry rift valley where the Minas Basin and Bay of Fundy now lie. Here, volcanic lava spewed from great fissures, sediment was dislodged from the surrounding highlands, and small dinosaurs roamed about.

THE HUMAN WORLD

It would seem that the first hamlet came into being at Minudie about three hundred years ago as Acadian French spilled over from the Chignecto Isthmus in the Amherst area. The Acadians built dykes and drained the marshlands of their new home. As the population of the Maritimes grew, settlers established an interior route via River Hebert and a long ridge known as Boer's Back. Above this route were lakes and a small river, which emptied into the area that is now Parrsboro. A ferry service connected the region to the Annapolis and Avon valleys, which were then the centre of Nova Scotia.

After the expulsion of the Acadians in 1755, new English-speaking settlers arrived. Only at Minudie did the Acadians rebuild, though many English settlers remained there as well. The new settlers were Yorkshire English in the valley of the Southampton-Maccan area, New Englanders at Parrsboro, and, after the American Revolution, large numbers of Loyalists along the Parrsborough Shore.

In the last half of the nineteenth century the area boomed. The large coal fields were developed and towns grew up at Springhill, River Hebert, and Joggins. A railway system was built linking these towns to the new Intercolonial Railway. One of the two major spur lines had its centre at Maccan, with trains reaching the main line west from Chignecto and east from Joggins. In addition, Joggins was developed as a port. The other spur went from Springhill to Parrsboro, from which huge amounts of coal were shipped. When the coal industry collapsed a century later, so did the spur lines.

On the Parrsborough Shore, the boom was brought about by shipbuilding. The coastline was dotted with shipyards taking advantage of the excellent timber of the area and the numerous coves. When the twentieth century brought the advent of steel ships, one by one the shipyards closed.

Between the Parrsborough Shore and the coal fields lay the eroded remains of the Appalachians and the interior valleys where farming and lumber mills were the mainstay. For a time, when water power was king, numerous dams were built and small factories and grist mills spawned fortunes. All these, too, disappeared in the twentieth century. So what is left? A beautiful area of high cliffs, forested hills, prosperous farms, and wonderful beaches. Three small urban centres remain: Springhill, Parrsboro, and River Hebert-Joggins. History tied the region together, and adversity has tested it. Today it waits—a land of promise.

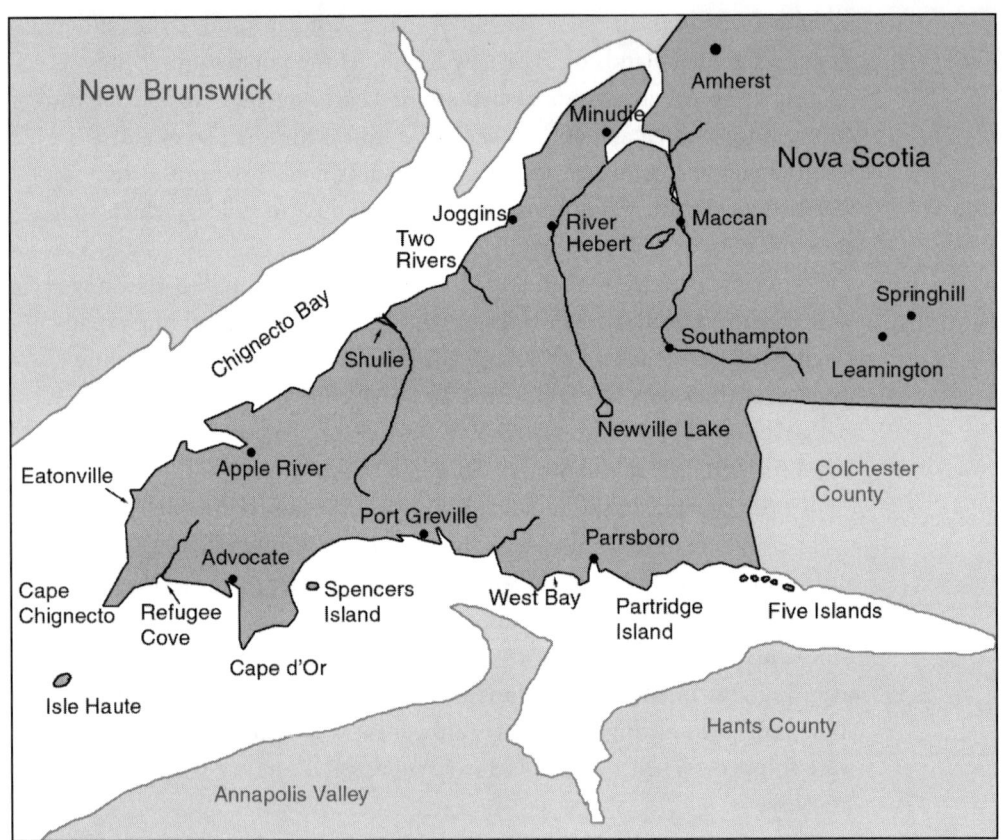

CUMBERLAND COUNTY SOUTH
(not to scale)

Chapter 1

The Coal Fields by the Sea

JOGGINS, RIVER HEBERT, MACCAN, CHIGNECTO

The Arseneau Pit, River Hebert, 1910

A pile of slag-working debris can be seen on the left at this small mine on the Joggins seam.

Joggins, River Hebert, Maccan, Chignecto

The village of Joggins is located on the Bay of Fundy's Chignecto Bay. The narrow coal seams so typical of this region originate beneath the nearby sea. The northernmost seam is Fundy; to its south lies Forty Brine; farther inland, the Kimberley seam begins. The Queen seam is also on the coast, and, finally, the southernmost is Joggins. Slightly to the north and underlying the Joggins seam is Hard Scrabble. Farther inland other seams appear, often joining the ones running from Joggins.

The coal seams in this field were usually narrow, ranging from four and a half feet (137 cm) at the original Joggins mine to about two to three feet (60–90 cm) in River Hebert and an impressive thirteen feet (400 cm) at Chignecto. The entire field stretches inland for about 20 miles (30 km) through Fenwick, ending at Styles Brook. This field would spawn an amazing total of 83 official mines and claim about 100 lives from 1869 to 1978. The coal found here has a higher sulphur content than that of the neighbouring Springhill Field, where 14 official mines were developed and 445 lives were lost.

In the eighteenth century, the French were presumably the first Europeans to notice coal protruding onto the shore. R. Morrow's book on the Springhill explosion, published in 1891, reminds us that the French were among the first in the area:

> [Joggins] had also evidently been used as a fort by the French or other soldiers, as cannon balls and other fragments of warfare were found imbedded in the earth near the edge of the bank above the mine, and there still remained [c.1847] earthworks constructed to the height of three or four feet.

In *Off Trail in Nova Scotia*, Will R. Bird states that Joggins coal was for sale in Boston by 1720, and that New England entrepreneurs were mining it. The log of a Captain Hale in Salem, Massachusetts, speaks of mining coal from Joggins cliffs in 1730. The first official mine on the shore was opened circa 1847, and would come to be known as Joggins No. 7. The owners, the General Mining Company of London, England, were reluctant to establish the mine—the mines in Pictou and Cape Breton counties were ample to satisfy demands. But, under attack by those who opposed the company's coal monopoly, the General Mining Company opened the Joggins mine in an effort to show that they did have the interests of the province in mind.

The company found that two English mining families were already in Joggins digging coal and selling it in Saint John. The spot was isolated; one could get there either by boat or via a small path from River Hebert. A third option was to follow the shore from Minudie. The company set about clearing land, building log homes, and starting their 50 ton-a-day production. In the early 1870s, the mining property was sold to some merchants in Saint John. It was soon divided in two, and the Hard Scrabble Mine came into existence.

As the nineteenth century progressed, Joggins No. 2 became one of the main mines, and here the mine's first roundhouse for trains was built. With the development of the Shore Mine (No. 7), the train line was extended and a new roundhouse was constructed. The Maccan Mine on the west side of the Maccan River was opened in 1858; River Hebert's Victoria Mine opened in 1859, followed by the Lawrence Mine (east side of River Hebert) in the same year. In 1863 came the Chignecto Mine, in 1864 Maccan's St. George, and in 1866 the Minudie (Kimberley). The coal era was off to a good start.

Getting the coal out was a challenge, however; it had to be transported by way of the tiny Joggins "harbour" or by vessels able to navigate the small Maccan and Hebert rivers. The solution came with the Intercolonial Railway, which opened in 1876. As a national line built away from the American border and eventually extending from Halifax to Montreal, the rail link permitted the Springhill- and Maccan-area mines to take advantage of a huge market. In 1883, a 12-mile (18 km) track to Joggins was begun. It was completed in 1887 and would last until 1961. In 1888 at River Hebert, near the Kimberley Mine, a new line was constructed to Minudie to take advantage of its wharf. Never very successful, this line was closed around 1910.

The mines supplied a range of jobs: pit timber suppliers, dock and ship workers, farmers, merchants, and railway workers. And then there were those who worked in the mines. At the top of the hierarchy were the mine managers. Next came the underground managers, who were responsible for cutting, loading, and removal of coal, as well as overseeing the ventilation system, the machinery, and worker safety. Next in the chain of command was the overman, who was responsible for a number of mine sections. Each section had a foreman called a shotfirer, who would prepare the dynamite used for loosening coal. Near the bottom of the hierarchy were the miners, the workers who picked, shovelled, put up side and roof timbers, and laid down tracks for coal cars. Other jobs included attaching and unhooking coal box cars, pipe fitting, and, as time went by, electrical work. In addition, drainage systems had to be set up. The water was collected in cisterns in the mine bottom and pumped up to the surface.

In the nineteenth century, children worked in the mines. Some were "trapper boys" who sat in small alcoves cut into the walls, alone and usually in the dark. Their task was to open and close a door that controlled ventilation and the passage of coal cars. The trappers could be as young as nine years old. Older boys drove carts pulled by horses or worked in the stables inside the mines, caring for horses or oxen. Under government regulation, the horses had to be brought above ground once a year for a break. Great care had to be taken so that the daylight did not blind the poor animals. Of course the advent of electricity meant horse power was no longer needed. In the small mines—and there were many—boys and men did the work of the horses as well.

Because Joggins No. 7 extended under the ocean, there was always the danger that the sea would break into the mine. In 1926 the mine was closed and the

Explosion, Victoria Coal Mine, 1930

The Victoria coal mine exploded on Sunday evening, September 17, 1930. The funeral for Wilfred White, William White, Emil Kralicek, and Simon Fowler was held at St. Denis Roman Catholic Church in Minudie the following Sunday.

bankhead torn down. A new mine opened a bit further inland at Maple Leaf No. 4. Accidents and death were unavoidable in the mines, but the 1930s brought three major disasters.

The first tragedy struck at River Hebert's Victoria No. 2 mine in 1930, the first year of the Depression. At 6:40 in the evening on September 17, a night crew was working in 1500 East, No. 1 balance. (A balance is a large tunnel off a main underground level from which smaller diggings, called bords, are excavated.) Twenty men were in the mine when the explosion came.

On the surface, rocks flew out of the mine and up the track toward the bankhead complex. The ground trembled, but in most of the town of River Hebert the tremble was so slight it was scarcely noticed. Almost immediately, a volunteer group went down the slope, though they lacked proper equipment. They soon encountered a heavy, extremely potent gas known as after damp, which results from the original gas explosion. They met survivors crawling toward the open air and sighted the bodies of the dead. Weakened by the fumes, the rescuers were forced to withdraw toward the surface, where others had to drag them out.

Meanwhile, draeger rescue equipment was being rushed from Springhill; ambulances and undertakers were summoned from Amherst. Families, friends, and relatives crowded around the pit head. For hours they hoped. First came the survivors, and then the dead. One haunting image was that of Henry White, a miner, who went down the slope and reappeared carrying the body of his son William, only 21 years old. In total, seven men died, and three others were badly burned. The dead were as follows: Philip Brine, aged 64, with a wife and two children; William Burke, 52, with a wife and one child; Wilfred White, 47, a widower with three children; Simon Fowler, 45, with a wife and three children; William White, 21, with a wife and child; and Clarence McGraw and Emile Krawlick, both 21 and unmarried.

What had happened? An odourless, tasteless, colourless gas known as fire-damp had built up in the mine. Lighter than air, fire-damp builds up first on the mine ceiling. All day they'd had trouble with the booster fan that was used to clear gases. It had been down several times, for a total of at least one hour and five minutes. The danger was recognized and the fan was repaired. Once it started, however, it probably drew in air from abandoned workings through small bore holes. The men had been told to dim or blow out their lights, but then they returned to their work. There was no underground manager on duty.

Eight months later, in the spring of 1931, an explosion occurred at River Hebert, Victoria Mine No. 4 (called the Bright Light). The mine was about a year old. On May 11, forty men were working on the coal face; the shift had not been long at work. At 8:15 in the morning, George Quinn fired a shot to bring down a coal shelf, and suddenly the gas exploded. Miners were knocked to the floor and underground coal cars were thrown from their tracks. Tons of coal fell, blocking the exit. In the mine depths, the workers recovered from the daze and gathered together. Upon discovering that the mine-level exit was blocked, they entered the airway tunnel. People rushed to the pit head as smoke rolled from the mine.

At first it was feared that all might have been lost, but within an hour, the majority of the crew had reached the surface. They brought with them the names of the dead: Adolph LeBlanc, Sanford Legere, Charles Stevens, Sam Rector, and George Quinn. Shortly after eleven o'clock the last body was recovered; it was that of Thomas Jones, who left a wife and ten children. On May 12, the *Amherst Daily News* quoted a Mr. Muise as saying: "The coal industry working two or three days a week cannot exist under present conditions any more than the average individual can." The article goes on to say that after deductions some men returned to their families with pay envelopes containing as little as 19 cents. The mine was permitted to flood, which extinguished the fire. Some weeks later it was opened for inspection; eventually it reopened. On June 3, 1932, the people of the area unveiled the Miners' Monument.

In 1932, disaster struck Joggins. Twenty-two miners were working at the Old Maple Leaf Mine, No. 4 (nicknamed the "5 and 10") on December 2. The work tunnels were squat—only 19 to 22 inches between floor and ceiling. The miners, noticing a break in the belt of the booster fan, sent for William Hachey to repair it. The crew continued loading the coal onto conveyor pans headed to the main haulage as Mr. Hachey finished the repairs. At three o'clock in the morning, a gas explosion took the lives of five men. They were found in their work positions, killed instantly by the blast and flames. Volunteers poured in. "Not heroism, simply duty," said one miner. The dead were: Ezra Murray, a well known ball player for River Hebert; Charles LeBlance of Amherst; William Hachey; Dan Boudreau; and Henry LeBlanc. As the decades went by, the mines closed one by one: Maple Leaf in 1939, Victoria No. 4 in 1948, Last Jubilee in 1951, Bay View in 1961. The last mine was the Cochrane, which closed around 1980.

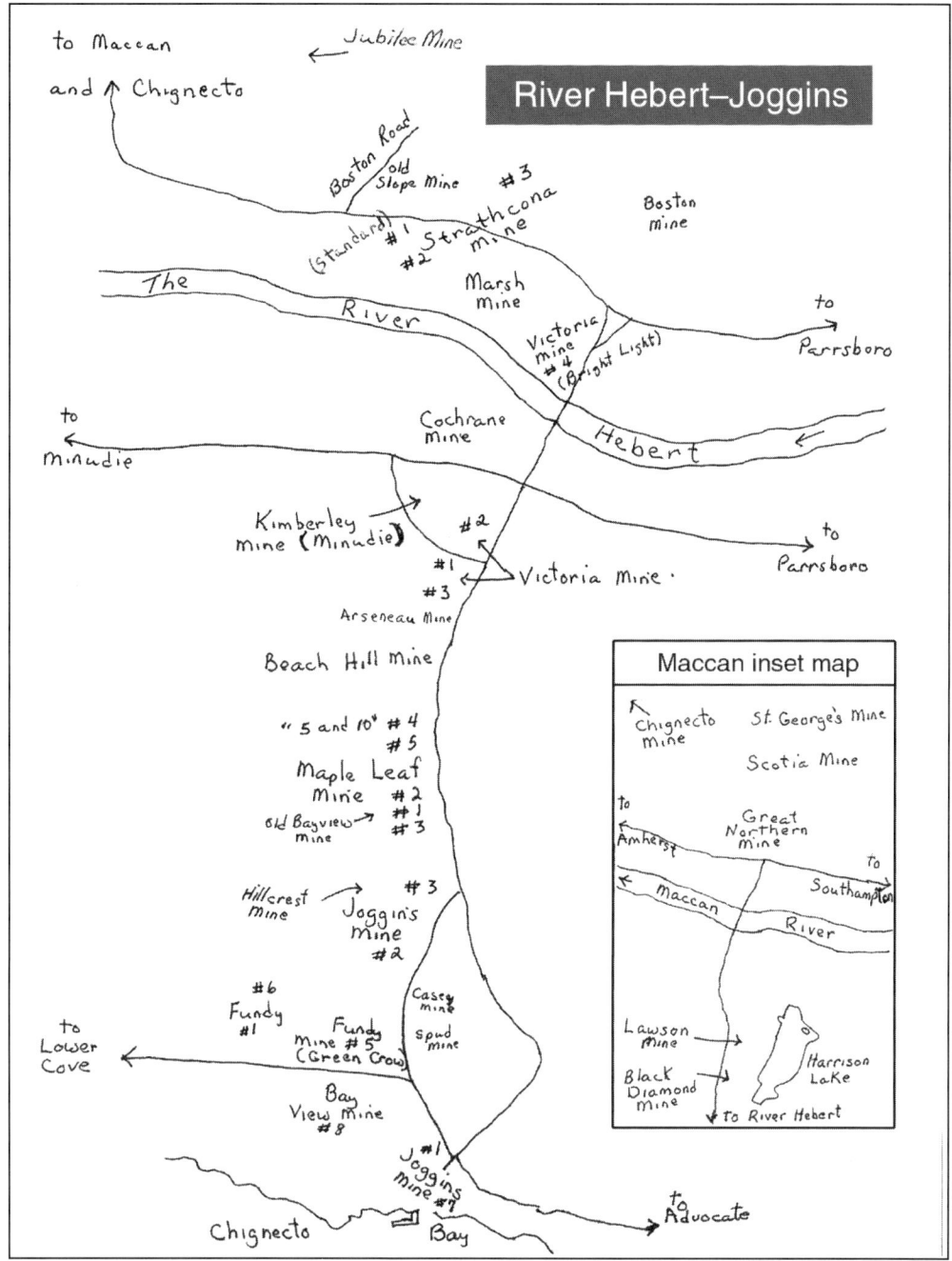

THE MAJOR COAL MINES OF THE JOGGINS/RIVER HEBERT/CHIGNECTO FIELD
(not to scale)

THE COAL FIELDS BY THE SEA

FIVE MINERS

The following descriptions of the lives and deaths of five miners are representative of Cumberland County's miners.

JAMES ARTHUR REID

James Arthur Reid was an African-Canadian born near the town of Guysborough, Nova Scotia, in 1881. His birth name is unknown, but by 1901, he and his younger brother John were adopted by John and Eliza Reid. In 1908, he wed Daisy in New Glasgow and went to work in the Thorburn mines (likely the Marsh Mine). By 1910, he was working in Joggins No. 7. One of his jobs was to crawl through the shaft that removed any water that seeped down to that elevation or that had been pumped up to the underground drain level. His task was to make sure the passage was clear to the cliff edge. Later, he went to the Maple Leaf Mine No. 4. He and Daisy had eleven children. From the *News and Sentinel*:

> [James Reid] and James Fowler were finishing the machine cut on the wall [of the 5 & 10] when stone fell from the roof and pinned Reid to the pavement. Five men were required to remove the rock. He was taken home and on Monday to the Amherst hospital. His hip was very seriously injured and it will be some time before he recovers.

It took a year for James to recuperate, but he never fully recovered. He did, however, work above ground at Fundy Mine, sifting through the waste stone for coal before dumping the stone over the cliff. Whatever coal was found belonged to those who found it. By this time, James's son Don had quit school to help his father at the cliff work. A man named Jury supplied the horse, so they split their earnings three ways: one bag of coal for Jury, one for his horse, and the third for the Reids. From there, James went to the Bayview No. 8. He retired in 1950, when he was about seventy years old, and died five years later.

THOMAS WHITE

THOMAS WHITE

Thomas White was a man of great spirit and determination. He'd lost both legs just below the knees in a train accident in Amherst. He took up the trade of sharpening scissors and knives, and eventually worked at the mines. His job was to empty the rock chute; any waste coal he found was his. Thomas built his own box-shaped "boots" that fitted onto the stubs of his legs. He had his own horse and wagon. Some people remember that his horse always stopped for a drink at the pit pond on the way home. One day, the horse went right into the middle of the pond, leaving Thomas stranded until help arrived. He hunted and fished and danced at parties, and appeared happy. He died during the 1960s.

WILLIAM HURLEY

On January 1, 1913, five or six men had been working at the Lawson pit across the Maccan River from the station. William Hurley was left there in the evening to keep the boiler fired. A tube in the boiler burst and a stream of scalding water was thrown over William, killing him. A bachelor, he had lived in Maccan, and had formerly been employed in Jones Livery Stable, Amherst.

FIDELE D. LANDRY

Fidele D. Landry, a World War One veteran, was killed by a fall of stone at Maple Leaf Mine No. 4 in early spring, 1938. He left behind a wife and three sons. According to a newspaper account, after the "long cortege of World War veterans and miners marched from the residence to St Thomas" Aquinas church- for the funeral; "the remains were then taken to Amherst" for burial.

ROY TIPPING

At noon on January 24, 1940, at the Bright Light Mine (Victoria No. 4), Roy Tipping was thrown from a mine car, striking an iron bar in the fall. Hours later, he was dead. The mine carpenter, Mr. Tipping had entered the mine to do some repair work. When the coupling of the car he was in broke, the car ran backwards down the slope. Fred Johnson, in another car, caught his fellow miner. Tipping suffered a fractured skull and ribs, as well as internal injuries. He left a wife and six children.

THE PIT CART

After being hauled above ground by the pit cart, the coal passed through a screening process. Men sorting the coal dumped the good coal down the coal chute. The rest, which went down the stone chute, was hauled away, dumped, and hand-sorted. Good coal was thrown into rectangular ton boxes. From there, it was shovelled onto waiting vehicles and sold. These workers were not paid by the mining company; their wages came from the sale of waste coal. Both James Reid and Thomas White oversaw men working for them at the stone chute; each worked a different mine.

RIVER HEBERT STREET SCENE, C.1900

RIVER HEBERT — The River Hebert, with vessels at anchor, and the railway bridge to the right. The *Chignecto Post,* July 23, 1885, reported: "There are fourteen vessels in the river this morning loading with coal and deals [lumber]. Messrs. Young have loaded 5 or 6 within a week with deals for Parrsboro."

S.S. HARBINGER — The waterfront on the River Hebert, just below the railway station on Station Street. The vessel is the S.S.*Harbinger*: it was used as a tug boat to escort vessels on the river, and it also travelled to Joggins and Saint John.

**MACCAN-
JOGGINS
LINE**

The train arrives at the station, River Hebert.

**WILF AND
ALVAH
CARTER'S
CHURCH**

Baptist Church and stable, c.1910, River Hebert. During World War One, its minister was Henry Carter.

THE MILL The Thomas Rufus Christie Brothers Mill, River Hebert Head, c.1910. This is the area of the present Chignecto Game Sanctuary.

VICTORIA MINE NO. 1

The Victoria Mine No. 1, 1914, in River Hebert showing the bankhead in the centre and a large cliff on the right.

THE RIVER HEBERT MINERS' MONUMENT, UNVEILED IN 1932

A full list of the dead is found in Appendix A.

THE SHAMROCKS

Small-town life during the Depression in River Hebert. Around 1936, the girls won the Nova Scotia Women's Softball Championship in Truro. They practised at the school and at a ball park near the present legion. They had no money and no uniforms; for travel costs, they depended on donations. Janet (Skinner) Wilson remembers that Blondie, her 13-year-old sister, was the youngest player. When the older ones went out to celebrate, Janet and Blondie had to stay behind with their chaperon, Cecil Morris, who had promised the girls' parents that they'd be well supervised. Their father worked in the mines, and their grandfather ran a store in Beech Hill.

The Shamrock team members were (from left to right):

Back Row: Fred Johnstone, Dena Burke, Jessie Rae, Pearl Carde, Dorothy Rae, Cecil Morris, Yvonne Burke
Second Row: Lottie MacArthur, Jannet Skinner, Kate Burke, Marjorie Melanson, Minnie Kaye, Martha Wolfe
Front Row: Agnes Gates, Geneva Wood, Catherine (Blondie) Skinner, Neta White

THE LAST COAL MINE The River Hebert Mine (the Cochrane) in a 1972 snowscape. This was the last mine in the area.

TRAIN WRECK, JULY 12, 1919 A train bound for an Orangeman festival was wrecked at Maccan when the bridge fell. The engine and tender got over safely, but the middle car dropped into the river. The weight drew the locomotive back and it became suspended.

Train wreck, c.1944

This wreck occurred when swaying pulp cars caused the passenger car to upset. Pictured are Robbie Bezanson, Wilfred Burbine, and Clarence Fullerton.

THE LIGHTHOUSE AT THE END OF JOGGINS DOCK, C.1947

JOGGINS PORT The old Joggins Wharf in the 1950s. The ruins of the tramway from mine No. 7 is visible to the right of the wharf.

THE PARADE
DOWN MAIN
STREET,
JOGGINS,
JULY 1,
1920

On December 31, 1928, the Joggins streetscape changed forever when an explosion occurred in the basement of the Wonderland Theatre, where coal had been burning to heat the structure. Local entrepreneur Fred Burke owned the building, which also served as his home. (Burke's son Edmund gained fame for receiving the ill-fated message about the *Titanic* in 1912.) The fire spread quickly, and there was no fire department in Joggins. Nor was there a water hydrant system, so when the Springhill and Amherst fire departments arrived, they found their equipment was useless. Water had to be thrown on the fire with buckets. When the fire was finally extinguished, the school, theatre, two hotels, and many shops, homes, and barns had been destroyed. The town never completely recovered.

MAIN
STREET,
JOGGINS

Main Street, Joggins, looking east in the early 1900s.

THE COAL FIELDS BY THE SEA

MINE NO. 2   Going into Joggins pit of Mine No. 2 in 1900.

MINE NO. 7 The No. 7 pit, Joggins, 1918, with the rail line in the foreground. Notice the covered-in roof to the mouth of the pit.

MINE NO. 7, The new bankhead at Joggins mine No. 7, located by the sea.
NEW
BANKHEAD

MacCarrons River Bridge, Joggins

The abutments are still visible today. Near this bridge was the pumping station, which supplied water to two ponds used by the mines.

The story of the Joggins to Maccan rail link (1887–1961) is not without its tragedies.

Around midnight on April 10, 1911, Charles Burke, age 14, was returning with friends from a horse show. At Maccan he hopped aboard the baggage car of the Special, which the coal company was running to take fair visitors to River Hebert and Joggins. A mile and a half from the Maccan station, Charles decided to move to the passenger car. He slipped and fell to the tracks: "the wheels of the baggage car passing over his body and severing his right leg from his body, cutting off his right arm at the elbow, and also cutting off the fingers of the left hand" (*Amherst News and Sentinel*). A doctor on the train did all he could as they backed the train to Maccan, but the boy lingered for only a few hours. The special train then went on to Joggins; the boy's adoptive parents were notified and brought by train back to Maccan.

In early February, 1917, 30-year-old Dr. William W. Herdman took the afternoon Maccan train to Strathcona to deliver a child. He was in River Hebert to replace Dr. Munro, who had joined the army. Further along the line, at Jubilee, a winter storm raged so badly it was decided to back the train to Joggins. Unfortunately, Dr. Herdman was walking back on the tracks and was run over by the returning train. They found him on the tracks the next morning. He was buried in River Hebert.

THE NUMBER TEN Number Ten Engine, Maritime Coal, Railway and Power Company Joggins-Maccan line.

JOGGINS PORT   Loading coal below Joggins Mine No. 7. The ship, owned by Joggins merchants, is the *Ladysmith*.

THE TERMINAL, JOGGINS

The Round House (locomotive shed and machine shop) in Joggins, around 1945.

JOGGINS WHARF Joggins Wharf c.1890.

THE TWO PIERS Joggins, 1894. A tug loads coal from the coal shute on the loading dock, while on the left several vessels lie by the wharf.

JOGGINS WAR MONUMENT

In World War One, River Hebert lost 19 men and Joggins lost 21. Their names are engraved "lest we forget" on two war monuments in Joggins and River Hebert.

A large number of men from the area joined the fight including Alex Seaman; Lyle Pugsley of Barronsfield; Leo St. Peter of Maccan; William Porter of Joggins; Robert Hatherly of River Hebert; and Leo and Alvah Carter of River Hebert.

SERGEANT ALVAH ELDON CARTER

Alvah Carter was killed in the muck of Passchendaele, Belgium, where a rather pointless battle was fought in flooded fields whose drainage ditches lay damaged from shells. Across it were boardwalks for soldiers to march on; horses or soldiers that fell off these boardwalks ran the risk of being sucked in and drowned in the muck. "Good God! Did we really send men to fight in that?" said one Britisher.

Professor David Beatty, a well-known Sackville historian, recorded Leo's story about Alvah in The *Amherst Citizen* (1984): "Alvah had just received a pair of hightop boots from home," said Leo. "There was friendly joshing about them and the mud." The boots, it seems, were sent over by his River Hebert girlfriend.

"I saw him, when they dragged him out. He still had those boots on. That was the only way you could recognize him." Leo himself would lose an arm to a sniper during the battle.

The news brought grief to River Hebert. In addition to his family, he left behind a girlfriend, Maud Greenfield. Likely, he never knew of her last gift to him—a cake she made and sent over from River Hebert. Maud later wed and had several children, but she still visited the Carter family in Pointe de Bute, New Brunswick. These words he'd sent to her:

Tho' afar from you I wander
I send this loving greeting.
May God have you in His keeping
Till we meet again.

In the photograph Alvah is holding his walking stick with a shell casing at its end. The family still has it.

Sergeant A.E. Carter
85th Battalion, [Nova Scotia Highlanders]
Canadian Infantry
30th October, 1917 age 21
Faithful to God, King and Country

Among Henry and Rose Carter's five sons and three daughters was Wilfred, born on December 18, 1904, in Port Hilford, Nova Scotia, and destined to be an international country and western singing star. Henry Carter was a Baptist minister who during World War One was based in River Hebert. Locals in River Hebert say that Wilf and his father did not get along, and Wilf left home when just a teenager after a dispute with his father. From then on he was making his own path, working around Cumberland County, and at the age of seventeen in Massachusetts, then returning to the Maritimes.

By this time the family had left River Hebert since the parsonage there had burnt in 1918. In 1919 Henry Carter began preaching in Pointe de Bute and remained there until his death in 1928. Rose Carter died in 1953.

WILF CARTER AND HIS DAUGHTERS, SHEILA AND CAROL, c.1951

Wilf had a distinctive yodeling sound: "I first heard yodeling by a Swiss fellow in a road show," he said. "At an early age I headed for the box cars and landed in the west where I got a job on a ranch." He was soon singing in the bunkhouse. His first radio broadcast was in Calgary in 1930, his first recording in Montreal in 1933. He became very popular. In America he was called Montana Slim. He died in Arizona on December 5, 1996.

At 2:20 A.M. on the morning of April 15, 1912, the *Titanic* sank on its maiden voyage several hundred miles off the coast of Nova Scotia. Meanwhile, in Joggins, 19-year-old Edmund Burke and his friend Fidel Ouellette were enjoying their favourite hobby: listening to the Cape Cod, Massachusetts, wireless station on their home-made receiving set. They took turns listening to the nightly broadcast of ship news on their headphones. After the eleven o'clock news, Ouellette went home and Burke continued to listen. He soon heard the Cunard liner *Carpathia* as it rushed to the scene, and the news from the *Virginian*. He heard survivors were being taken to New York. Edmund recorded the events on a blackboard placed in the window of his father's Wonderland Theatre. When local people read the bulletin, many did not believe it. They felt it was a lie that would make Joggins look silly. Some people even followed Edmund in the streets, hurling insults and threats. According to Harry Burke, County Councillor J.L. McDowell arrived and told the crowd they could be arrested. Edmund showed McDowell his wireless machine, and the councillor received confirmation from Western Union. The leaders of the local coal company arrived to check news on survivors.

On Monday, April 15, the *Amherst Daily News* carried a front-page story headlined "Steamer Titanic Struck Iceburg." There was no mention of it sinking or of Edmund Burke. During the following days, news of the *Titanic* filled the paper.

Burke's family was from a bordering village called Lower Cove. As a young man Edmund had worked in the Joggins Mine No. 7. In the 1890s, he had also worked as a mechanic in the United States. News of his *Titanic* message reached Halifax, and several days later his accomplishment was recognized by the coal company and the townsfolk. Edmund Burke went on to live a long life in the Joggins area.

EDMUND BURKE, 79, IN 1971

DONALD REID: KEEPER OF THE CLIFFS

Donald Reid worked at the coal mines for 21 years. In 1961, with the closure of the last Joggins mine, he found himself unemployed for the first time in his life. Donald found work in various parts of the province to support his wife and seven children. He had taken up fossil collecting as a hobby, and when he retired from the Nappan Experimental Farm, it became his full-time passion. Reid still works at the Joggins Fossil Centre in a partnership with his sons and daughter, passing his knowledge of fossils onto the many tourists that visit this world-renowned village. The Nova Scotia Department of Natural Resources has bestowed on Donald Reid the title "Keeper of the Cliffs" in recognition of his work. Pictured is one of Donald's displays.

TWO RIVERS: JULY 1888

Two Rivers today is an abandoned stretch of coastline about five miles (eight kilometres) from Joggins, on the way to Advocate Harbour. Though desolate, it is beautiful. In 1888, it was the centre of attention in Cumberland County. The photograph shows the launch of the Giant Timber Raft—a titanic cigar-shaped structure 592 feet in length, 52 feet wide, and 35 feet high tapering to 10 feet at the ends. It weighed an estimated 11,000 tons—the second heaviest object ever to have been launched up to that time. It held perhaps 24,000 logs.

Its purpose was to transport tree timber to New York City where it was used in myriad ways: constructing piers, pilings for sheds, ship staging, and slipways in vessel construction. The owner, James D. Leary, had insured it for $30,000. Though he was confident, success was not a foregone conclusion. Towing would cost a fair bit, but it was far cheaper than transporting all those logs on scores of ships and paying tariffs on the more finished lumber.

The timber raft arrived on August 8, 1888, towed by two steam tugs. This type of raft was designed by a Saint John lumberman, Hugh Robertson, who believed it was possible to transport logs by ocean direct to their market. It was the third such raft to have been built. The first was constructed in 1886. The *Chignecto Post* on Thursday, December 24, 1885, carried the following item:

> One enterprising townsman, B.B. Barnhill, Esq., of Two Rivers has commenced a large raft of timber to be built on the bank, cigar shaped, lashed

securely with chains and to be towed by steamer to New York. If this enterprise proves successful, it will revolutionize the timber carrying trade from these shores. The raft contains three million feet.

Huge numbers of people flocked to Two Rivers to see the launch on a Saturday morning at the end of July. They flooded the road with buggies, wagons, foot traffic, and horses. In a festive atmosphere, the women of the River Hebert Baptist Church sold food from a large shed (complete with tables) constructed just for this occasion. Photographers were also present to capture the moment—which never occurred. The raft wouldn't leave its slip to enter the sea. The next day they tried again; it went halfway into the water and there it stayed. Other attempts to float it failed, and it was eventually dismantled. Disgusted, the *Chignecto Post* reported the following on December 8, 1886:

> The public having wasted a good deal of shoe leather, horseflesh and bad rum on the "big Raft" are at present rather nauseated with it, but, a rumour says, it is to be rebuilt with a million more logs in it and to be launched again next spring.

THE GIANT RAFT The launch at Two Rivers, Cumberland County, July, 1888.

In 1887 a new raft was built—at a cost of thousands of more trees. This one was owned by James Leary. It was lost in a storm on Sunday morning, December 18, 1887, when it broke away (or was cut loose by a frightened crew) from its tug boat en route to New York. The *Chignecto Post* on Thursday, December 29, 1887, carried the following article under the banner "The Big Raft Wrecked; no longer a menace to navigation":

> ...dispatch from New London, Conn., dated Dec. 25 says:—The United States steamer Enterprise, Captain McCalla, arrived this afternoon after a successful search for the big lumber raft. When 125 miles south east by south from Nantucket shoals she fell in with an immense number of logs from the raft, proving beyond doubt that the raft had been broken up. The steamer was among the logs all day, and part of Friday night. None of the logs were lashed together, but were floating singly and spread as they travelled with the tide. The logs were not in the line of ocean steamships. Captain McCalla fears no damage from that source...

PORT GREVILLE, c.1904

The Elderkin Raft under construction.

Despite the success of the third great raft, no more were built until 1904, when a smaller one was constructed at Port Greville on the Parrsborough Shore, by William Elderkin and Company. Smaller is a relative term—it was still 460 feet long and had over 7,000 logs. The chartered tug that arrived from America on a Saturday was expensive, perhaps one hundred dollars a day. The highest tide was the next day and it was time to sail. Elderkin was a very religious man; after some thought, he decided that they would not sail on Sunday. As it turned out, however, Monday's tide in the unpredictable bay was fine. They launched and the raft was towed successfully on a nine-day trip to Boston. Elderkin believed that their good luck was a sign that God was pleased.

The photographs give a good idea how the 1888 raft was built. The logs had been laid in a construction cradle, and the timber intertwined in such a way that no joint would exist to weaken and sever the raft. The tapering logs were laid so that the top and butt ends were reversed, further binding it together. Small hardwoods were placed at right angles across each layer of timber to prevent shifting and to form a base for the next layer. Chains were threaded through in as straight a line as possible to fasten to the lengthways centre chain. Finally, numerous chains were wrapped around the outside of the bundle and fastened to the side chains. This was expensive, but the plan was to reuse the chains in other rafts.

Not everyone was saddened by the demise of the rafts. This innovation was not popular among local mill owners who feared that the export of logs (not rough-sawn lumber and planks) would eat into their businesses. Ship owners also stood to lose, as they would not be transporting nearly as much timber to America. And, of course, it would have meant much less employment in Cumberland.

Chignecto today is a small village, just to the east of Maccan, through which a paved road meanders to Fenwick. Its population is less than 100, and it shows little evidence of its historic importance. In 1911, however, Chignecto had a population of 684; by 1921 it was 843. What happened? The coal mines of the area closed.

There were several mines here: Chignecto No. 1, No. 2, No. 3 No. 4 and No. 5 as well as nearby mines such as St. George. In addition, the Maritime Coal, Railway and Power Company (owners of Standard Engineering) had been the first to put into practice Thomas Edison's idea of "pit head power." In theory, this meant that power could be transported more cheaply by wires than by railway coal cars. In practice, it meant that the company put in a small plant to generate 500 horsepower at the mouth of a Chignecto mine. They expanded again and again. In 1911, power lines were run all the way to Joggins, making its mine the first in Canada to be completely operated by electricity. But Chignecto lacked an adequate water supply, and in the 1920s the plant moved to Harrison Lake, Maccan. With the mines and the power plant gone, Chignecto lost its reason for being.

KING COAL MAKES ELECTRICITY

Construction at the Maccan power plant, 1931. In the background is Harrison Lake, the damming of which made the lake larger.

CHIGNECTO MINES

THE BOILER ROOM IN CHIGNECTO MINES, C.1900

THE COMPANY STORE, CHIGNECTO, C.1900

A TRAIN IN CHIGNECTO, C.1900

Chapter 2

The Elysian Fields

MINUDIE

THE LITTLE GIANT

The Little Giant—a threshing machine at Baron Baker's barn in Barronsfield—photographed on the day a new engine arrived from Ohio. The thresher was owned by Alfred DesBarres. Pictured are Louis Terrio, Norman DesBarres, Fred Burke (on the pole), Baron Baker, Douglas Baker, Kenneth Baker, Alfred DesBarres, and Matthew Smith.

Minudie

The story of Minudie is one of promise and tragedy. Minudie lies at the mouth of the River Hebert, five miles (eight kilometres) from the village named for that river. Directly opposite is the other major river of this area: the Maccan. This is the beginning of the Cumberland Basin, a tide-swept area whose estuaries churn the brownish mixture of mud-laden waters and send a large wave—a tidal bore—up both rivers for miles. Three kilometres away, on the other side of the channel, the rise of the Beaubassin (Fort Lawrence) Ridge begins. Between the channel and Minudie proper is a marsh, ten kilometres square, that today is a government-owned common land.

The Mi'kmaq frequented the area, drawn by the food supply: herring, smelt, gaspereau, and salmon in the spring; shad, mackerel, cod, and flatfish in summer; eels, salmon, and trout in the autumn; smelt and seals in the winter. There was also a supply of clams, mussels, and other shellfish. It was a fisherman's paradise. In addition, there were moose and caribou, and many plants to harvest and berries to pick. Here Mi'kmaw women constructed wigwams—dome-shaped, wood-framed dwellings covered with birchbark. When it was time to move on, the wigwam was torn down and transported. By the time of the arrival of the European settlers, disease had reduced the Mi'kmaw population by perhaps 75 per cent.

Around 1672 Jacques Bourgeois, a surgeon and fur trader, sold his farm in what is now the Annapolis Valley and founded a settlement near present-day Amherst called Bourgeois colony. In 1676 Michel le Neuf was granted all lands between Tatamagouche and Petitcodiac. He arrived four years later, but was instructed not to disturb the settlers already at Bourgeois. Le Neuf called his land Beaubassin. It grew quickly. The 1686 census shows a population of 110; eventually settlement spilled over into the Minudie area. Dykes were built and the fertile marsh reclaimed. Among the early celebrations held by the people was "Bird Day," which took place on Monday the week after Easter. The people went to church to pray that birds would not descend on their harvest as they had once, causing a winter of hunger.

The village grew through the years. The calendar marked the British Conquest in 1710; the return of Louisbourg to France in 1748; the building by the French of Fort Beausejour, a mere five kilometres across the channel and up a hill, in 1749. In the same year, the British founded Halifax as the new Nova Scotian capital.

One of the driving forces behind the subsequent decision to expel the Acadians was the Battle of Grand Pré, which took place in the Annapolis Valley in the darkness of February 11, 1747. Colonel Arthur Nobel and his command of 470 New Englanders, billeted in that village, were taken by surprise in the early hours of the morning by a force numbering less than 300. Most of this force came from Quebec, although some were Indians. Having snowshoed from Beaubassin, they overpowered the sentries and, using axes, began breaking into

MINUDIE SHAD

A late nineteenth-century view of Minudie fishing shacks located near the dykes. These structures were built in the early part of the 1800s. There were two rows of sheds with a wagon trail through the centre. Men can be seen cleaning and splitting shad. The hamlet is well known for its shad fishery.

THE SKYLINE, MINUDIE

Minudie Village c.1900 as seen from the direction of the rivers. The house just to the right of the churches was the Vernon House, which a family member was to ship to Ontario.

the houses. Nobel, still in his night clothes, was cut down. His two brothers and 74 others were killed, many in their beds. Sixty-nine more were taken prisoner, but the remainder fought on and were permitted, upon surrender, to return to Annapolis Royal. Some Acadians had aided the attackers.

By April, 1750, war clouds loomed. A British fleet landed, bringing Colonel Charles Lawrence and 400 men. Their purpose was to build a fort on the British side of the Missaquash River. A fanatical priest, Abbé Louis Le Loutre, had learnt of the British plans. Churning up a panic, he had his followers, mainly Indians, burned Beaubassin and the surrounding area—much to the despair of the villagers. The reason for this action was to deprive Lawrence of the labour and supply base needed to build a fort; the Acadians would then be forced to live on the west side of the Missaquash River in what was claimed as French territory. The Abbé himself burnt the church. Lawrence left, only to return in September to build his fort. In 1753, Lawrence, who by now had no use for French Neutrals, became governor of Nova Scotia.

On June 16, 1755, a 2,000-man New England force captured Fort Beausejour, renaming it Fort Cumberland. In early July, an army lead by Commander Braddock, chief of all British forces in North America, marched into the Ohio country of the Pennsylvanian wilderness and was subsequently shot to pieces by the French and Indians. Braddock was an impressive figure. Four horses had been shot from under him; he was mortally wounded as his fifth horse went down.

During these events, the Minudie French had tried to stay completely neutral. Nevertheless, 30 families from Menoudy (Hebert, Landry, Foret...) as well as 23 from Nanpan and eight from Meccan had gone to the French side of the river. Colonel Monckton, the British commander of Fort Cumberland (the former Fort Beausejour), sent a company of New England volunteers led by the Irish lieutenant Dixson to destroy Minudie. They arrived at night, forming a line at the back of the village. At sunrise, with a discharge of muskets, they advanced. The people fled from their homes. Because the tide was in, flight across the mud flats to Amherst Point was impossible. Many jumped into the water and were shot by the troops, who watched their corpses float away down the river.

There were many other examples of barbarous behaviour. On the banks of the Nappan River, which flows into the Maccan opposite Minudie, a group of Colonel Goreham's rangers came upon four starving Acadians, resting unaware by a dyke after a search for food. The officers turned their backs; the Acadians were instantly shot and scalped. The scalps were passed off as Indian at Fort Cumberland—because Indian scalps brought a sizeable bounty. Some officers were bitterly opposed to this policy.

One officer who would dominate the Minudie area was Joseph DesBarres. DesBarres was born in May 1729 in Montbeliard, France, to a French Protestant family. In 1765, this land speculator was granted the Minudie estate—8,000 acres of valuable Acadian land, part of it called the Elysian Fields. Many Acadians were returning from exile at that time, but DesBarres wanted English tenants. When he tried to evict the Acadians, they simply moved elsewhere on his lands.

(After all, it had been their land only a decade or so earlier.) He gave up trying and left the area in disgust, but went on to have a successful political career, eventually becoming lieutenant-governor of Cape Breton and later governor of Prince Edward Island. DesBarres was also a skilled coastal map maker and became well known for the Atlantic Neptune, a detailed chart of the east coast of North America. He lived until 1824. It was said he could dance a jig on his 100th birthday—though he died at 95! He is buried in Halifax.

Among the area's resident Acadians was one Zedore Gould, born in 1727. As a youth of 20, he had aided the French in their raid on Noble at Grand Pre. During the Expulsion he escaped to the Miramichi country. Ironically, he returned to become a tenant of Governor DesBarres at Minudie. Another Acadian name that survives from this period is that of Peter Brine. He went into hiding in the forest during the Expulsion, remaining there for years. While in hiding, he developed skills in astronomy and weather predicting. Upon surfacing, he became known on the DesBarres estate as a human almanac.

A 1795 report shows that each tenant family received 200 acres, to be held by them and their heirs forever. In return, they paid a rent in animals and crops. DesBarres built the mills, receiving one quarter of the toll. Most—if not all—of the original tenants were French, with such surnames as Foret, Burbine, Leger, Melanson, and Como. Graham wrote in 1795 that Minudie had "twenty or thirty houses and a chapel" and that one could "cross a river in a log canoe" or a boat belonging to a nearby family named Glennie "upon which you enter the township of Amherst." The ferry service operated until 1929 when the *Little Dan* ceased to operate and ended up rotting on the Nappan River. Another estate, belonging to Colonel Barron, bordered that of DesBarres. In time it would become the village of Barronsfield.

On a moonlit night in September 1796, an eight-year-old runaway stepped ashore—shoeless—at Minudie/Barronsfield from a log canoe. For reasons unknown, he had left his home at Lower Maccan where his parents lived. He was lucky. He was taken in by the Brines who kept him overnight. Some time later, his father came and got him; time would bring him back. In 1846 he wrote: "It is just fifty years since I first sat my bare feet in the Minudie Shore, for I had no shoes" (*King Seaman Diary*).

His name was Amos Peck Seaman, born January 14, 1788. He later dropped his middle name:

> I was born at the John Peck House at Sackville, N. B. a hut under the hill…I had an uncle by the name of Amos Thomas [as] a young man came from the Un. States a brother of my mother he took the Small pox died and was buried under a [poplar] tree near the house. I can Jest Remember him. for him I was named…[as for] Peck when I came to write my name I did not like [it] & so left it out the word Peck appeared small to me.

Amos's father was from Wales, his mother from an established Welsh family in America. They arrived in Sackville from New England in 1764, part of a wave of

KING SEAMAN'S HOME, BUILT 1836, AS SEEN IN THE LATE 1800S

settlers who came to replace the Acadians. Here, in poverty, they tried to raise their eleven children. Nathan, the father, would sit by the water and write poetry, musing over his non-conformist religious beliefs, while Zena tried her best to make ends meet. She attempted to teach her children to read, using a Bible, but when little Amos reached Minudie he was, in fact, illiterate.

Amos Seaman was hard working and frugal. In his mid-twenties, he worked during the day and went to school in Maccan at night. Perhaps this is how he met his future wife, James Metcalf's daughter Jane, whom Amos married on May 12, 1814. He loved his "Jinny" all his life:

> 50 years ago this very day I took to my Self a Wife & with her I have lived a Happy but Buisey life that's a fact
> Amos

In 1814, it was recommended he get a grant of land, and two years later he built his first store. In 1818, he bought half an acre of land at Minudie. His wife taught him to read. One of her great joys was gardening. They would have eleven children.

> five Living sons and four daughters to day all at their old hom Sorrowing for their Sick Brother William now on a sick Bed fromwhich all fear he may not Recover. James [grandson] to is verry low & wasting away
> January 1, 1860

> February 4th at 8 P.M. poor Williams Spirit left the Body which was laid away on the 7th alls Will son James left famely and friends on the 16 day of May following William.

> poor [son] Job Sick & Suffering dropsy Set in Cant be long with us.
> January 14, 1861

> all is well Suffering over & Job at rest
> is laid away famely troubles Boys behaving
> Bad 2 only left
> June 13, 1861

These two sons were Gilbert (d. 1896) and Rufus (d. 1888), and they would outlive him. He'd lost five sons: Ephraim (1842), Amos Thomas (1856), William (1860), James (1860), and Job (1861). His daughters were Ann (MacFarlane, d. 1902), Mary (Vernon, d. 1889), Jane (Hibbard, d. 1904), and Sarah (Mitchel, d. 1915). The only one to die as a child was Ephraim; on his grave at Minudie is the following epitaph:

> Oct. 22, 1837 aged 5
> See Israel gentle shephard stand
> With all engaging charms
> Hark, how he calls the tender lambs
> And folds them in his arms.

By the mid-1820s, Seaman was collecting rents for the DesBarres. The next step was to get the lease for the quarries—which he did. Grindstones were an important part of the grist mill industry. They were used for sharpening knives and axes, and for polishing metals. Seaman soon found himself involved in legal disputes because he got grindstones from the areas exposed when the tide went out. Most of his legal problems ended when he purchased the DesBarres estate in two deals: one in 1833, the other in 1840. The price for the first deal (which excluded the manor farm) was over 8,500 pounds.

The grindstones often came from high-quality hard sandstone in the nearby village of Lower Cove on Chignecto Bay. As the tide lowered, workers rushed to this exposed bedrock and split away slabs of it, first using gunpowder to dislodge it from the underlying rock. Roughly cut into a disc shape, smaller ones were taken to shore. For the large stones, an iron pin was wedged into a central hole, and the stone was then lifted by a boat at high tide and floated to shore. These stones could be as large as six feet in diameter. A nearby land quarry also produced high-quality stones. In 1847, over 33,000 grindstones were exported at a tremendous profit. Seaman also owned the ships to transport the stones. In 1843, he opened Nova Scotia's first steam-powered mill:

> on Thursday last…Steam Mill first set in motion, so that this day may be called the anniversary of the Minudie Steam Mill, the first ever in the Province. I have built this mill at an expense of 1500 pounds. Bring in your logs 30,000 or more. She will cut them all up before the year is o'er
> Sunday, August 13, 1843

The mill also produced flour.

Seaman was well known for using rhymes, and employed them in reference to a tragedy tied in with his mill. People came from all around to see the mill, including four young men from Amherst Point. They never made it; the boat capsized. One made it to shore, another kept himself afloat by using two spruce plough handles.

> The Lord He praised this scrubby spruce,
> Was never made without its use
> It picked me up upon the sea
> And saved my life with passage free
> While two sweet youths both young and brave
> Were doomed to have a watery grave.
> July 4, 1843

In 1846, Seaman made a trip to England and was presented at Queen Victoria's court. He had come far from the poverty of his boyhood.

And Seaman was generous with Minudie. A strong believer in education, he made sure his children had the education he had been denied in his youth. In the 1840s, he built the schoolhouse that houses the King Seaman Museum. He also erected the Protestant church and paid one quarter (400 pounds) of the cost for St. Denis Roman Catholic cChurch. In his own religious beliefs, like his father, he was a thinker and a seeker. Though he would welcome the Anglican bishop Inglis, he himself was a Universalist (today the Unitarian Universalist Church), a tolerant group that does not accept the trinity and believes truths about God can be found in both the Bible and other sources. Perhaps this explains some of Seaman's thoughts:

> We can then Sing with the dying Christian
> farewell, farewell, to all below
> My Jesus calls and I must go
> Then Launch my Boat upon the Sea
> this land is not the land for me
> Praise be to God Our hope on high
> The Angels sing and so will I
> Where Sarups bow and bend the knee
> O, thats the Land the land for me"
> 1843
> "Thanks be to him who rules on high
> who governs all below the sky...
> it is Right that I Shud Suffer
> it is so ordered for my good
> Sunday, March 3, 1844

His advice to his son Gilbert when the youth went to Boston in 1840 was "to be deligent & to have communion with few…be intimate with One, deal justly with all and speak Evil of none." Though proud, he was also compassionate: a hard luck story could bring him near tears. In short, he was both a great and a good man.

His marriage was difficult at times because Jane developed mental problems. She outlived him by two years, dying at the age of 73 on April 23, 1866. On her tomb is the following epitaph:

> Oh though aft depressed and lonely
> All my fears are laid aside
> If I but remember only
> Such as those have lived and died

AMOS SEAMAN, 1788–1864

The Grindstone King of Minudie. He was an entrepreneur whose imposing store, lands, dock, ships, and public buildings made him truly a local king.

**JANE
SEAMAN—
1793–1866**

Jane (Metcalf) Seaman died in 1866 in her 73rd year.

Seaman foresaw some of the problems that would plague Minudie after his death:

> My eyes are dun my head is White and I can scerely Git a nap at Night poor deluded Sons Why could you not have waited a little Longer before you demand a division of Property which for the last 50 years of my life I have worked and toild for as but few others Ever did.
> September 18, 1861

> I am yet a Striving hard to Save the Ship.
> January, 1864

Seaman and his horse, inspecting the estate, had become a part of folklore. Four months before he died he wrote:

> my Self & old Horse Charley ages together is near 102 me in my 77th and Charley in his 25th—we go the Rounds quite well together yet.
> May 12, 1864

Several days before he died came these words:

> When I am dead there lay me down beneath
> some lonely sod
> Let me return to dust again, the spirit
> to its God…
> Place there a lonely stone or not,
> it maters not to me
> Whether remembered or forgot this mortal
> man may be.

Amos Seaman died after a short illness in September 1864. *The British Colonist* said of him: "The poor will lose a benefactor and his deeds of kindness and hospitality will make his memory long cherished by all who knew him." To satisfy his wish to be buried by water, a small lake was made. In it were placed water lilies from England. They are still there. His tombstone bears the words: "The affectionate and beloved father/Though dead he yet speaketh."

The next hundred years in that area would belong to coal and the settlements of River Hebert and Joggins, while Minudie slid further and further into decline. Seaman himself was interested in coal as early as the 1840s, and had been part of the pressure group that led to the opening of Joggins Mine. Technological change and poor management of his estate meant that his grindstone business, his hay and cattle export empire, his ships, wharf, and ferry would all disappear.

On October 4, 1869, a severe storm—The Saxby Gale—drove water levels six feet (2 metres) above any previously recorded at Moncton. The dykes at Minudie's Elysian Fields burst and the sea poured into the marsh.

As years went by, Seaman's handsome home, Grindstone Castle, was abandoned. The house that had entertained a who's who of society was occupied by others, stood derelict for decades, then was gone. After Seaman's death there were attempts to continue development. In 1888, a railway line was put in from the Kimberly (Minudie) Mine in River Hebert to King Seaman's dock near the old home. Not very successful, it closed in 1910.

**KING
SEAMAN'S
LOWER COVE
GRINDSTONE
QUARRY, MID
1940S**

All that remained was the 40-foot-high smokestack of the steam plant that powered the lifting crane used for raising large stones. The three boys in the photo were the sons of Edmund Burke: Leo, Eugene and Arthur. Today, virtually all trace of it is gone save for deep-flooded excavations. On the shore at low tide, one can see many grindstones that were left because of flaws.

**THE WHARF
AT MINUDIE,
1920**

Captain Merriam's ship is loading lumber. Just visible is an ox team and horses.

MINUDIE DOCK, c.1925

On the right is Gerome LeBlanc's little store which was abandoned. LeBlanc is described as a very lonely man who never wed - a man with many sayings. One story is told of how he used to advertise in the papers for a woman. He could write a nice letter, they say. One day a lady arrived but was very disappointed by either his appearance or wealth or both. She went to King Seaman's grandson, Amos and said "Amos, you've got to get me out of here!" He hitched up the horses and she left. Harry Burke used to visit him. A teacher, Burke was well known for his pioneer work on the Joggins fossil cliffs and for his writings. His sister is the mother of the famous singer from Springhill, Anne Murray.

DR. JIM BALMENO AND HARRY BURKE

THE ELYSIAN FIELDS

THE MINUDIE RAILWAY ENGINE, 1910

MINUDIE TODAY King Seaman's three remaining public buildings. From left to right St. Denis Roman Catholic Church, the King Seaman School and the Protestant church (Universalist).

PLEADING LOVE LETTER OF JAMES METCALF, JUNIOR

James Metcalf was 65 years old when he, his wife, Mary, and one of his sons, James, landed in Maccan from Yorkshire. Over the years, more of his children arrived: Richard and William around 1790; Joseph (with his family) in 1820. The old man acquired land first in the Maccan area and then, in 1785, 500 more acres just two grants south of Southampton's present-day cemetery. I have abridged the following letter and modernized its wording. It was written by James Metcalf, Junior to Ann Gill, England, August 1772.

My Dear,

This letter is to let you know that I am in good health as I hope you are. We are many leagues apart but distance and length of time since we parted have not made me forget you. I have got 207 acres of land…a good part will be easily cleared because it has been done by the French…There is a little fly called the mosquito that is troublesome in summer time and bites like a midge but…it is becoming much fewer because of land clearing and grazing. It is the only thing I have to say against this country. And now you promised me that you would surely come and my dear I shall be very glad to see you keep this promise and…if you come I will be a kind husband to you…I must marry for I cannot live well as I am now…Don't be afraid of the voyage. I suppose you will have many from Yorkshire with you. I wish I were the lines of this letter so I could be your companion. But I have much work to do and cattle to look after. I pray to our God to support and protect you. When I came over I was sick two-and-a-half days after we left England and again when the sea was very rough. But we had a good passage and were healthy.

The people here are mainly Presbyterians and Baptists. The Anglicans are a smaller group. The people here are naturally kind to one another; even the Indians when one comes to their wigwams will give you meat if they have any. Spinning wheels are very expensive. We use English money and a thing called a dollar. If you come please bring a bushel of wheat—four different types. Keep it from salt water, lay it like a pillow in your bed. Bring some tea in case you are seasick. Sometimes we make smoke at our doorways in the evenings to keep the mosquitoes out—they are more troublesome than you can imagine.

If you come you will sail up to Fort Cumberland and once you are there write me a line or two and send it to me on the Maccan River and I'll pay the man who brings it and come for you. Let me know by letter if you are coming or not. You pay the passage at Liverpool but if you haven't got the money make friends with some others who are coming and I will pay them. Write to James Shanks at Liverpool about it.

I will conclude for this time. May the Lord bless you and bring you safely here.

James Metcalf (junior)

P.S. If you write to me direct the letter to Maccan, near Fort Cumberland in care of [Lieutenant] Governor Francklin at Halifax.

As there is no mention of snow in the letter, it seems he wrote it shortly after he arrived. When Ann Gill received his letter, she set out. She arrived at Fort Cumberland in 1773 or 1774. As instructed, she sent a messenger to Metcalf who was awakened at two in the morning. He started at once for the Fort, and they were wed that day. Metcalfe leased his farm land from Michael Francklin—a part of the vast Francklin Manor. They had no sons but five daughters. Elizabeth, born in in 1778, married the younger Charles Atkinson; Mary lived from 1783 to 1787; Hannah was born in 1787 and died shortly afterward; Ann, born in 1789, who wed William Sharpe; and Jane, born in 1794. Jane would marry Amos "King" Seaman of Minudie on May 12, 1814.

James Metcalf, Senior, died on his Maccan farm on March 13, 1787. He was 80 years old. He was buried on the farm. By this time, the family was leasing land from DesBarres. McDonald, a DesBarres agent, wrote of the younger Metcalf in 1795: "This man…does well; he appears to be a modest man and an intelligent farmer."

In the late 1790's Metcalf sold their land in Southampton to Thomas Harrison for 18 pounds. His will, with an estate worth 3,880 pounds, was probated on January 1, 1820. He left:

> grandsons, James Metcalfe Atkinson and Amos Atkinson, sons of daughter Elizabeth and her husband Charles Atkinson all…homestead farm, 520 acres, eighty of which is marsh, all to be divided equally with my stock and utensile

The young man from Yorkshire did well.

Surprisingly, Seaman's mother was against his choice of bride. Also surprising is James Metcalf's decision to leave all his property to the children of one daughter and none to Seaman's sons, Amos Thomas (born 1815) and James (born 1816). (And the will was signed on October 17, 1817.) Census reports only to add to the confusion: in one, Hannah and Jane were both born in 1787!

There is a story that Elizabeth Metcalf had eloped with her first cousin when she was sixteen years old (1794). They were wed and she became pregnant. However, the minister was unqualified, so the marriage was annulled. The groom returned to England. Elizabeth bore a girl, named Jane, who was raised by the Metcalfs (especially Ann Gill) as if she was their own daughter.

An elderly lady whose passion is Minudie filled in the rest of the story: "Things seem to have gone well for James Metcalf until a church service in Lower Maccan, when a visiting English minister recognized him. Then there was trouble. James had left a wife and five children behind in England; his marriage to Ann Gill was annulled and she left him. She then married Charles Atkinson and had more children. That also might explain why he sold land in Southampton—he needed the money."

Chapter 3

The Fertile Inland Valleys

SOUTHAMPTON AND NEWVILLE

NEWVILLE, C.1905

A view of Newville showing its mill on the water with a huge smoke stack near it, a large store, and railway tracks—a hive of commerce. On the lake to the left is the log boom, where logs were held until they could be sawed. The mill burned in 1912. In the background is the village of Halfway River.

SOUTHAMPTON AND NEWVILLE

In the upper part of the Maccan River valley lies the village of Southampton. Once a part of Maccan settlement, by an act of the province of Nova Scotia on April 18, 1872, the village reverted to the name on the original land grant awarded on February 15, 1785. Each of the 20 Yorkshire families received 500 acres. Some of the families had been settled on the land for several years, and the grant gave them official recognition.

Two hundred years ago, where the west and east brooks meet the Maccan River, Henry and Elizabeth Furlong ran an inn that supplied rum to the locals and was a popular meeting place. Henry, a Roman Catholic, had married Elizabeth Harrison, a strong Methodist. After Francis Boss died, the couple purchased the property from Boss's son John, but Mrs. Boss had not signed over her share and it took years to settle with her. Near their home the Furlong Bridge was built. Across the stream a second home was built—perhaps for their son, James. When the time came, Henry and his wife were buried near the stable and near the resting place of Francis Boss. Today the road runs over their graves, and the Central Hotel, built by William Harrison, stands where the old Furlong house once stood. The older brick structure was reputedly the scene of a fire that killed the family of John Harrison.

The *Acadian Recorder* of Saturday, April 16, 1825, records the "melancholy accident," which took place on April 4, 1825:

> On the night of the fourth, the dwelling house of Mr. John Harrison, Junior (aged 40) of Maccan (Southampton), Cumberland county, was consumed by fire, and shocking to relate, Mr. Harrison and his three children perished in the flames. The only person that escaped the devouring element, was a female relative, who, after being nearly suffocated in a fruitless attempt to rescue the girl that slept with her, broke through the small window of her apartment into the street. The flames immediately burst through the window and made it impossible to render any assistance to the child, the sound of whose voice twice vibrated in her ear. This young woman was sleeping with her head covered, and the bed clothes had taken fire when she was awoke by Mr. Harrison, who, it appears on awaking, called to her to save herself; and rushed out of his room, whether in order to gain the outer door or with what view is left to conjecture. His remains were found about three paces from his bedroom door and near where the fire must have originated.
>
> About three weeks previous to this melancholy accident, Mr. Harrison had consigned his beloved partner (Elizabth née Cannon) to her tomb and his children were deprived of her maternal care. Now they are gone the way appointed for all living. "Be ye also ready."

The date of construction of the Central Hotel is unknown but William Harrison got ownership of the property in 1844 from James Furlong, who was in debt.

Eventually ownership passed to his son Matthew, who died of injuries he sustained when his arm got caught in a threshing machine. His widow married John Siddall.

Below the Central Hotel a large dam was erected by William Humphrey c.1820. There he had both a grist mill and a sawmill. The top of the dam became a well-trodden path for the locals. In 1850 Humphrey sold the dam to a William Adams. From the map it is easy to see how the river was the lifeblood of the early settlement, determining its homes, roads and industries.

One problem they encountered was black bears. These animals raided their sheep and cattle in the daytime, forcing them to place their domestic stock in strong enclosures at night. On Harrison Hill (granted to John, senior, though occupied by his son) the second John buried his wife.

At that time, it was common practice to lay loved ones to rest on the family farm, but the bears tore away the earth. To keep the animals from digging up a body, very large logs were placed over the grave. Finally, in the 1820s, the village established a cemetery on Harrison Hill that still serves as a burial ground. The first grave was that of John Harrison the third's wife. This young man, grandson of John, Senior, lived near the Southampton crossing of the river. Unfortunately, on April 4, 1825, three weeks after his wife's death, he and their three children perished in the flames of their burning home.

When the roads were built, Southampton was at the obvious position for river crossing. A log Methodist church had been erected on Harrison Hill about 1819. The minister, William Smith (1782–1842), lived beside it in the mission house. In his book *The Clockmaker*, written in the early 1830s, Thomas Haliburton says:

> The road from Amherst to Parrsboro is tedious and uninteresting. In places it makes so straight that you can see several miles of it before you, which produces an appearance of interminable length while the stunted growth of the spruce and birch trees bespeak a cold thin soil…Here and there occurs a little valley with its meandering stream and fertile soil…strikes the traveller as superior…from the contrast to the surrounding country.

Near or at what was to be the site of the Central Hotel, Sam Slick, with his Yankee charm, found himself involved in the case of "Fire in the Dairy." It was the home, he says, of Squire Bill Blakes. While he talked to the frugal old lady, the teenaged black servant, Beck, swept up the place. Finally, tea was put on. Haliburton describes the scene when Beck announces there is a fire:

> Oh Missus! Oh Missus! There's fire in the dairy! I'll give it [to] you for that—you good for nothing hussy. Your carelessness: Go and put it out this minute. How on earth did it get there?…put it out and save the milk. I [meaning Sam Slick] am dreadful afeard of fire, I always was from a boy and seeing the poor foolish critter seize a broom in her fright…

Sam Slick grabbed the tea kettle and rushed after the servant girl, following her into the blackness and collided with her:

I heard a splash and a groan and I smelt something plague sour, but I couldn't see nothing. At last I got ahold of her for she didn't scream, but made a strange kind of choking noise and by this time up come Marm Blake with a light…poor Beck…she had gone head first into the swill-tub and the tea kettle had scalded her feet. She kept a-dancin' right up and down like one raven distracted mad and boo-hooed like anything.

Later, as Beck tried to clear her hair, the old lady railed at her while Slick "haw hawed right out. 'Good gracious, marm, you forgot the fire.' 'No, I don't,' said she. 'I see him.'" And seizing the broom that had fallen, she went after the dog whose name was Fire: "'I'll teach you to drink milk. I'll larn you to steal from the dairy!'"

Communication between the old world and the new was kept up. The following are excerpts from letters between the Southampton area and the Malton area in Yorkshire, England.

Maccan River,
June 24, 1810

Dear Cousin,

Long ago I have had it in agitation of writing to you…hoping these few lines will find you and your family [well]…I settled here on this river [with his father William and the other children] about 23 years ago [1787] upon lands which had never been cultivated, all a wilderness. We cut down the wood of the land and burned it off and sowed it with wheat and rye, so that we have made…a good living. Here we make our own sugar, our own soap and candles. We spin and weave our own linen and wool…

I settled here… we all drew 500 acres of land each. I bought 500 acres ajoining mine which cost me about eighteen pounds…

We milk ten cows, keep one yoke of oxen, three horses and…twenty to thirty sheep….We generally kill every fall six or eight hogs…

The disadvantages we have is…the Winters being so long. There is six months to fodder our cattle, and, what is worse, the snow…sometimes four feet. The past three or four Winters have been very moderate….We have very much trouble with bears, as they destroy our sheep and cattle.

John Harrison

I have two sons…Pray send out a ship load of young women, for there is a great call for them that can card and spin. The wages are from five to six shillings a week.

March 14, 1817
Malton, England

A nephew of mine…proposes visiting…you…I might visit you again…

You will perhaps have heard what an extraordinary wet summer the last was, more so than the oldest person can ever remember. The whole of the

> wheat crop sprouted…parts so bad that it was not fit for anything but giving to the hogs…the farmers are emigrating by thousands…
>
> David Smith

The year 1816 was to go down in history as the year without a summer. The snows of late spring were brown on both sides of the Atlantic. The sun rose each morning as if through a cloud of smoke: red, rayless and with little warmth. Frosts hit each month through the growing season. The crops failed. The reason for the strange weather, though unknown at the time, was the eruption in 1815 of Tambora Volcano in the East Indies.

In the early days the destitute were auctioned off as servants and labourers, as the following accounts from the township records of Southampton show:

1822: Meeting held at Mr. Henry Furlong's at eleven o'clock in the forenoon on the first Monday of July. Agreed that this township doth agree to keep one half of Henry G., pauper for life providing that the township of Amherst should agree to keep the other half…during his life; if they do not we shall stand trial…

1827: Henry G., pauper, being put up for auction. Martin Hoeg bid him off at £17, 18 shillings for which sum he is to furnish the…pauper with sufficient meat, drink, washing, lodging, clothing and tobacco for the term of one year from this date and to return him at the expiration of the sid date as well clothed as he received him…

1833: Anthony S., pauper, put up for auction. Taken for the sum of £5, 18 shillings by Martin Hoeg for one year.

1834: Funeral expenses of (Anthony S.), pauper. To Martin Hoeg have the sum of £1, 10 shillings for funeral expenses of Henry, pauper.

1837: Jerimiah N., pauper put up to auction, to be provided for by Martin Hoeg for £14, 15 shillings. Two children (John Edmund R.) and Amilia L.) put up for auction. Lowerst bidder Andrew Herrett for one shilling and ten pence per week.

The industrial potential of the river was recognized early. Around 1788, about a decade before he died, Robert Ripley of Maccan built the first grist mill on a tributary in the Lumley grant. It was the first of numerous industrial buildings, for water power was king in the late eighteenth and much of the nineteenth centuries. There were problems with water power on a small river as the *Chignecto Post* notes on August, 1885, in reference to Southampton:

> Mr. L Tucker is about to erect a grist mill in place of one destroyed by flood…most of the mills hereabouts are closed for want of water.

THE SOUTHAMPTON WOOLLEN MILL, BEFORE IT BURNED IN THE MID-1880S.

Around 1871, the Southampton Woollen Mill was built by a joint stock company headed by J.T. Smith. It was located at the entrance to what came to be called (after the arrival of the railway in 1877) the Station Road. Its source of power came from damming the river and bringing the water to the mill by means of an artificial channel. As seen in the photograph, it was a substantial building, five storeys high and 100 feet long. It was never very profitable. To augment their incomes, many farmers raised sheep in this upland region, but the wool produced was coarse. This restricted the products produced, though in 1882, for example, it produced 1,500 blankets. Ownership changed twice; to Davison and Sons in 1877, and to Amos Atkinson in 1880 with C. Burton Seaman as manager. Then it burned. On October 29, 1885, the *Chignecto Post* (Sackville)—in addition to reporting the collapse of the Riel Rebellion in western Canada—ran this article:

> Matters remain in statu[s] quo as regards the woollen mill. There is yet nothing done towards rebuilding and rumours are very uncertain. The shop, formerly run in connection with the factory (it still stands) has been closed. Mr. Lusby's is now the sole mercantile establishment in the place.

The year 1886 was a year of hope. Christian revival meetings swept the region. On January 28, 1886, the paper reported:

Mr. Travis, manager of the Southampton Woollen Mills, passed through here on Tuesday to see the people of the district with a view to rebuilding. Should the people of that place offer sufficient encouragement, the company will begin operations at once and put up an improved factory.

As far as can be ascertained, however, the new factory was a smaller building. In 1903, it became the Valley Woollen Mills with O.B. Schurman as proprietor (the home he lived in can be seen in the back left centre of the picture). The factory finally closed around 1920.

It was replaced by the Nova Scotia Hardwoods Product Company, which ran what the locals called the "handle factory." From the picture, it seems that the factory occupied the old woollen mills. It made broom, axe, and hammer handles, rolling pins, and other such products from the local hardwoods. A sluice was used to bring in timber by way of the Lawrence Brook, terminating at the Lawrence Railway Siding. The logs then went down the river to the dam. A log rail line was constructed to reach the mill; tram cars with concave wheels hauled the trees. These cars were pulled by a motorized "vehicle," driving on its rims on the log track. The factory lasted only a few years.

THE HANDLE FACTORY The Handle Factory that took over the second woollen mill and added the side structure, c.1900.

SOUTH-AMPTON FROM STATION ROAD, ABOUT 1925

On the right can be seen the second woollen will which had become a woodworking factory and seems to be abandoned. Close to it, by the Maccan river, stands Ed Bradshaw's Blacksmith Shop. In the centre foreground of the photo is the home of Ed and Sarah Bradshaw, behind it the Office and then the Second Empire style Valley Inn Hotel. Just beyond it can be seen the peak of the roof of the village store.

Taking advantage of the prosperity brought about by the woollen mills, George Davison opened a hotel opposite it—The Valley Inn. This beautiful Second Empire style home still stands. His original dwelling had burned down.

There were Mi'kmaq in the area until around 1930, when most were moved to the Shubenacadie Reserve. They tended to travel through the back country, staying in particular at an area known as Indian Hill, just beyond the Lawrence Siding. One born in Southampton was Levi Pictou. When his mother decided to have him baptised in the Methodist church (erected 1875 near the woollen mills), a local woman exerted much pressure to have the child named William Gladstone after the British prime minister, but at the last moment his mother went with the name she wanted—Levi. The Mi'kmaq hunted, fished, and sold baskets (some large enough to be used as cribs); as far as can be determined, their farming was restricted to the Franklin Manor Indian Reserve near Newville Lake.

The first store in the area had opened much earlier, likely in the 1870s. It was located slightly left of the woollen mills and just up the hill. "Colonel" Charles Lawrence had moved the structure down to the village, where it was to remain for a century. Lawrence died in 1887. The new owner was A.B. Lusby. The store was the meeting place for all, and its owner would be a major village figure. Subsequent owners were David Mitchell; Lew Bird (also manager of the Handle Factory in partnership with his brother, Will R. Bird); W.C. Ogilvie (who for the year 1936-37 turned it over to two Jewish Canadians, Smoski and Lipwich, and then got it back again); Bill Myers; Robert Doyle; and, finally, the Milbery family. Of all these owners one would become best known—Will R. Bird (1891–1984), who was born in East Mapleton. His mother, a widow, was poor. After the death of his brother Stephen in World War One, Will R. joined the army. While he and Lew ran the store he wrote his first works. In 1923, they gave up the business. Will R. was to become the county's most famous author.

Further down the river from the cemetery was the carriage factory. It opened in 1888 when A.S. Fillmore, who had worked in carriage factories in the United States, moved to the bordering hamlet known as South Athol. In 1893, he built 125 carriages and about 60 sleighs and trucks. Unfortunately, on November 8 of that year the factory burnt down. Though he had little insurance coverage, he quickly rebuilt and began operations again the following year. By 1910, he employed 16 men. The market, however, was shifting to the gasoline engine. After another fire around 1914, the company closed and Fillmore moved out west.

For most people, life was hard. Upland farmers, such as those in South Brook, would leave their homes in mid-summer for a yearly excursion across country down the Etter Road, through Scouterac village, to the marsh land on the lower Maccan River. There they cut and piled hay. In the autumn they returned, leaving South Brook around midnight, with oxen and wagons. Usually one ox would break a leg on the journey. That was both good and bad luck; it was butchered on the spot and supplied fresh meat.

Etter Crossing at Scouterac of the Springhill-Parrsboro Railway had two sets of tracks. Here, men cut pit timber and loaded it onto flat cars which were picked up as the train returned to Springhill after carrying its loads of coal to the Parrsboro docks. From the *Chignecto Post* in 1885:

> An attempt was made to wreck the regular train…last Thursday morning by placing sleepers [horizontal beams to support weight] across the track near Etter Siding. Fortunately no damage was done.

As the century progressed, industrial activity ceased, with the exception of lumber mills. Some were stationary mills, such as William J. Brown's, Hoeg's, and Harrison's. Others, like that owned by Ernest Brown and then his son Carl Brown, were portable mills that moved from Southampton to wherever in the region there was timber to saw. The source of power for these mills was no longer water, but boilers.

Life for a mill owner was hard work. Carl Brown and his wife Gladys, for example, started their day at five: Carl got the mill ready and the diesel engine started; Gladys prepared breakfast for the 20 or so men who worked in the mill. The mill camp was covered in tar paper, without electricity or running water. All cooking was done on a wood stove. Mid-morning was quarter-time lunch; then

A GENERAL VIEW OF THE BROWN MILL

dinner; then the afternoon lunch; then supper and another small meal before the men retired for the night. There was also a baby to look after. By 1960, a cook's wages were five dollars a day. Carl worked in the mill. After it shut down for the day, he spent time tightening conveyor belts and making other repairs.

The decline of industrial activity coincided with the decline—and in several cases, abandonment—of Appalachian villages. The area's relatively infertile soil could not give people the lifestyle that catalogues and radios were depicting. Canaan Mountain, Little Italy, Glasgow Mountain, Newville, Scouterac, Shulie, Sand River, Thundering Hill, Eatonville, and New Yarmouth all ceased to exist.

MILL LIFE   Mill fires were always a danger. Part of the mill burnt in 1948, leaving a wrecked engine and a lot of expense to rebuild.

Mr. and Mrs. Alexander Lepper, grandparents of mill owners Carl Brown, in 1942.

The baby in the bath—at the mill, the baby always got the tub water first.

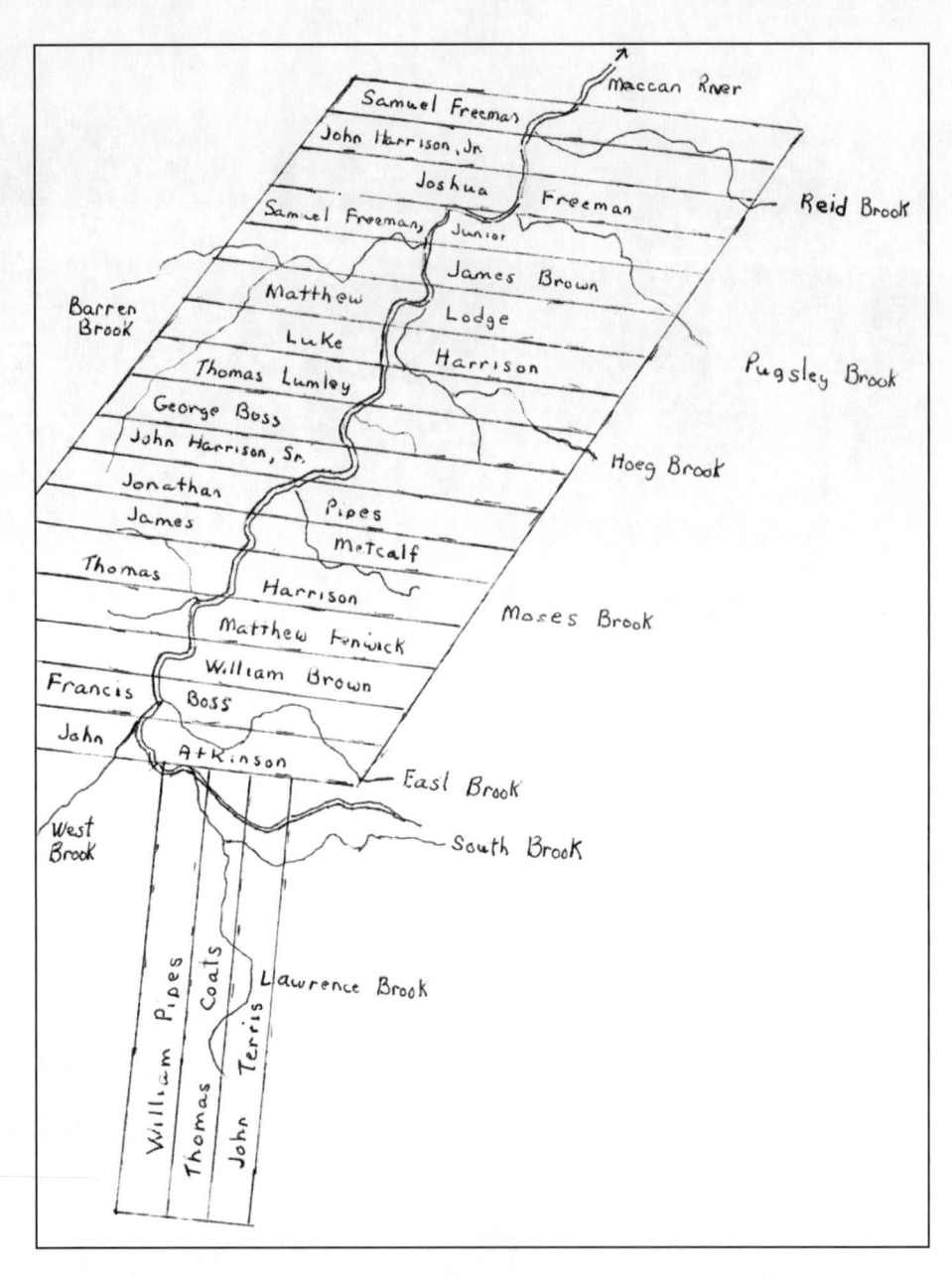

SOUTHAMPTON LAND GRANT MAP

NUMBER 12 The Number 12 Mail Train in Southampton, which ran on the Springhill-Parrsboro route.

WAITING FOR THE SPRINGHILL-PARRSBORO TRAIN

Waiting for the Springhill-Parrsboro train during the depression at the railway station in East Southampton are Sam Allen, mailman; at centre, Laura Goodwin and children and a friend. This station was on the left side of the road as one approached Springhill.

LEVI
BROWN'S
MAPLE
SUGAR
CAMP

A crowd gathers at Levi Brown's maple sugar camp to enjoy the first "sugaring off" of the season in 1909 in the village of East Mapleton. On the roof is Fred Hoeg, aged twenty. Seated on the right is Lina (Bird) Hoeg. The man with the hat is Levi Brown.

BROWN MILL, SOUTH ATHOL, 1939

Front Row (left to right): Ernest Brown, Jim Rolfe, Carman Bird, Elmer Siddall, Lawson Brown, Carl Brown. Back Row: (left to right): Bob MacAleese, Harold Bird, Lyorlus Brown, Eddie Hannah, Bill Atkinson

SPRINGHILL JUNCTION, 1949 Standing in front of the mill from left to right are: Clarence Gallagher (out of picture), Albert Fulton, Lawson Fife, Harold Henwood, Earl McLellan, Carl Brown, Leonard Brown, Will Hannah, John Small.

THE MAPLE SUGAR WOODS, 1909, IN MAPLETON

The Appalachian region of the county is well known for its hardwoods and the glorious autumn colours. In the spring the maple trees produce a lot of maple syrup. The photo pictures the Levi Brown sugar camp. Levi Brown is pouring the maple cream into molds held by Fred Hoeg. Martin Hoeg, his father, is standing to the left.

The Village Store

Evelyn (Gallagher) Brown remembers life at the Southampton General Store:

> It was 1944. I was working in Amherst at the old Metropolitan store on Church Street when I took sick and ended up in Springhill with my appendix out. Lyorlus, my future husband, didn't want me to leave Southampton again.
>
> Bill Myers ran the store in those days. 'Come down for a few days to work and see how you like it,' he said. That's how I got the job. I made three dollars a week. I would leave to have my family but in '54 I went back permanently.
>
> We were open six days a week—about 13 hours most days. If you wanted to smoke, you could. What I liked best about the job was the people. I remember French Mag who'd arrive from Lynn village in a horse and wagon, her money tied up in red and white handkerchiefs. She was a character. Cookies were kept in bulk in those days and Bill would give her one to keep her out of the bin. I also liked meeting the travellers. One of these salesmen from King Cole Tea gave me my first pair of nylons. I got married in them!
>
> One thing about Bill: he was a fine man. He was a professional; he wasn't a gossip. He'd been born in 1900, was of average height, ruddy complexion, and bald. Some people would run up bills they couldn't pay off. When they brought their order in Bill would say, as he packed them, 'We better begin the process of elimination,' removing raisins or whatever from the order. One day he was returning from Amherst when he came across an American woman with a flat tire. He fixed it and they fell in love. Her name was Margie and she was from Pittsburgh. They wed in 1950.
>
> We were busy. At the back in a room on the left was the post office. We'd even open it a couple of hours on Christmas afternoon. On the right was the storage room, the original Lawrence store. I fell in a barrel of sauerkraut in there one day; we sold it anyway.
>
> We also had the only gas tanks, the bus station, and we supplied the saw mills with everything from hardware to food. It was also the village meeting place. There were three of us who worked there: Bill, me, and George Manzer, who also did deliveries.
>
> Bill died so suddenly—Easter Sunday, 1957. He'd had little warning of an appendix attack. On Saturday I saw him leaning over the post office desk. "Don't tell Margie," he said. He asked me to sign a paper. I told him I should read it first. "Why?" he asked. "Because you told me when you hired me never to sign anything unless I read it." His reply. "So I did." It was his will. I signed it.
>
> "Buy something for Evelyn" he later told his wife. They operated on him in Amherst but his heart gave out. He was 57.
>
> I stayed on under the new owner Bob Doyle. By the time I left in 1965, I was making seventeen dollars a week. We moved to Oxford Junction and I got another job. It's funny how things work out.

SOUTHAMPTON, c.1954 The Knights of Pythias with the Pythian Sisters hold their late summer parade. Here it passes by Bill Myers' general store in the village centre. The store closed about 1980.

LAURA (FULTON) BROWN AND RAE (FULTON) DOHENY These two sisters moved, as many did, from Harrison Settlement to Massachusetts and married. Here, they are home for the summer and on their way to the Southampton General Store, August, 1956.

Newville Lake

One of the prettiest areas of this region is Newville Lake (once called Fullerton Lake), which covers an area two thirds of a square mile (one square kilometre). Through it flows the River Hebert. Both shores of the lake have their stories.

On the south shore were the small homes of the Mi'kmaq, a few miles away from their actual reserve, Francklin Manor, which is about three times the size of the lake. Their land was named for Michael Francklin, who came to Halifax from England in 1752. He opened a small rum shop, and with his education and integrity rose to become lieutenant-governor of the colony (1766–76). He was wise, popular, and well respected by both the Acadians and the Mi'kmaq. Like everyone else, he had problems with Governor Legge and lost his job. Around 1781, he became Superintendent of Indian Affairs, a post he held until his death the next year. He was also a major landowner in the Newville area.

Though the land on the reserve was good and there were some homes there, the Mi'kmaq preferred the lake shore where an earlier chief, John Logan, owned some land. They were a small population—about 77, according to the census. Ralph Thompson, reminiscing about the 1920s, spoke of Johnny Knockwood, a bachelor who lived with his mother in a home always filled with the smell of sweet grass; Levi Pictou, in his sixties, with a large family and relaxed lifestyle; Ben Knockwood, the hunter; Isaac Paul, the runner whose brother Denny was a heavyweight champion boxer in the army during the first war; Moses Francis, who worked at firing the boiler in the Harrison mill; Abram Glode, the Mohawk, from Maine, who ended up at Newville after he met a McGraw woman from Parrsboro who had gone to the States to work. The couple returned to Newville where Abram made beech pick handles for the Springhill Mines and was a good stepfather to McGraw's two children. Then there was old blind Sam Knockwood, who had to beg for food but was well liked by all. About 1930, most of the native people moved to Shubenacadie. The last, named Paul, left around 1950. Today there is no trace of them. Even their little chapel has disappeared.

The lake also had its tragedies. Around 1875, five young people went for a rowboat trip on the lake: Allison Davison and either his half sister or a Miss Barns, Clarence Fullerton, Miss Hatfield of Port Greville, who was teaching at Halfway River, and Miss Gabriel of Newville. The teacher playfully dabbled her hands over the side, tipping the boat a bit. Unfortunately, they all jumped to one side and the craft overturned. Davison, Hatfield, and Gabriel drowned. The other two clung to the side of the boat until rescued by Juby, a Mi'kmaq, who arrived by canoe.

On Christmas Eve, 1908, three youths from Newville went off to a skating party. As they skated on the Newville side, the ice gave way. Clifford Robinson and James Lockhart went into the cold water, while a third youth ran to the Newville Company store for help. Robinson's father rushed to aid his son; he, too, drowned. Their bodies were recovered Christmas Day.

The first settlers in the lake area were two Loyalists—John Fordyce and Caleb Lewis. They had escaped from the rebelling colonies. According to tradition, they both kept their whereabouts hidden from the Americans, and later on brought their families to the Newville-Halfway River area. Until 1840 this area was part of Kings County; the Cumberland Line was at the upper end of the Boar's Back Ridge (geologically termed an esker—a gravel bed formed by a river that once flowed inside a glacier). From there, the line cut across farms on the West Brook flat and into Canaan.

W.F. Jones built the first mill on the lake. Shortly thereafter, on its eastern shore, rose the first large Newville lumber mill. It was started about 1870 by a family named Young from Nebraska. It passed through several owners and eventually included company and private houses, as well as a store serviced by the train to Parrsboro from Springhill. The Newville Lumber Company's mill burned in 1905, but was rebuilt. In 1907, the 109-foot smokestack, still under construction, collapsed, and Horace Ledley was killed. In 1912 the mill burned again. It was not rebuilt. Today, nothing is left of Newville village except the name. The following lament was for the pioneers:

Aunt Peggy
(Margaret (Casey) Atkinson)

To new land farm in forest wide
Her partner lead her forth a bride.
He laboured on in sweat of brow
And forest fled from axe to plough
Together, they the harvest bound
And rejoiced to till the ground.
Wove table cloths and towelling
As white snow drifts on the wing.
She saved the poultery from the hawks
And herded bears from out the flocks,
No church in all the forest dim
The hoot of owl, her grandest hymn
No school to train the family young.
He who was hers for "Better, Worse"
Fell sick,
She saw her partner go before
Then stood alone on stormy shore
And many hopeful years did live
And many words of cheer did give.
But she is gone to peaceful rest
We took our now broad fertile lands
From early toiling settlers hands.
With reverence let us speak their names.

Samuel O'Donel Fulton*1829–1893
West Brook

Fulton was a well-known poet. He was a bright but delicate child. From the age of ten, when he contracted rheumatic fever, he was lame. His parents sent him to Horton Academy, and he later got a job teaching school in West Brook and in Mapleton. Fulton had a great speaking voice and was known for his work with the Methodist church and his stand on temperance. These topics got him invited to many meetings. Illness forced his retirement. Since there was no pension, he lived the rest of his life with family members. His father, Josiah, had acquired a second grant of land in New Prospect; after his death and burial in Thundering Hill, the family moved to the new farm. Fulton was a newspaper correspondent, and also wrote a temperance tale called *Red Tarn*, which was set in New Prospect (or Cascade Valley, as he called it). Published in 1873, it failed financially. A tarn is a lake; in this case, a lake he saw in a vision, a fearsome place, fed and swollen by the produce of stills (poison presses); a place of shadows, serpents, screaming birds, fighting beasts, and the ghosts of misery: "and over all the ghost DISEASE did clap, His fevered hands enraptured at his power." Fulton was admired for his conversation, piety, humility, and his acceptance of chronic physical pain. He died in Canaan and was buried in West Brook.

THE FULTON HOMESTEAD, c.1910 The new (Josiah) Fulton family home at Hidden Falls, New Prospect, c.1910.

ALLISON DAVISON

Allison Davison drowned in Newville Lake. Ironically, he was known as a great swimmer.

GLADYS FULTON

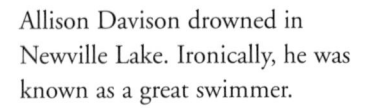

Pictured at left is Gladys Fulton, age 15, of Thundering Hill and her horse, Gyp, by the old Kings-Cumberland county line at the Henry Atkinson farm in West Brook, on June 10, 1930. Her job was housekeeper for relatives. Her mother was poor, her father deceased. Like many Maritimers, she had gone to Massachusetts—in her case to live with Laura, her sister, for several years. By the time she returned, her mother had remarried; but it was not a happy family, so Gladys entered the work force.

In 1936, the west part of the Atkinson farm was bought by the Department of National Defence as a site for a future emergency airport. The West Brook flat was about halfway between Halifax and Moncton, and at the time the plan made sense. The government paid $5,000 for 102 acres. They told the Atkinsons, in effect, that they could either sell it freely or have it taken from them. Though the family kept 200 acres, they moved to Truemanville in the northern part of the county. The airport was never built. The only local major development in that Depression decade was the establishment of the Chignecto Game Sanctuary.

Chapter 4

Where Dinosaurs Walked

PARRSBORO

PARRSBORO — MAIN THOROUGHFARE

Parrsboro, looking up the main thoroughfare toward the bandstand, which can be seen at the head of the street. On the right foreground was a triangular commercial block seperated from the post office (erected 1913) by an alley. This triangle-shaped building contained three stores: a barber shop, a hardware store, and maybe a millinery shop. This building was cut into at least two parts, which were moved to side streets and turned into residences. The war monument was erected in its place.

Parrsboro

At the base of the gap through the Cobequids is a small river named Farrells, which flows into the Bay of Fundy's Minas Basin. On the west side of the harbour entrance is a high island called Partridge Island. This area is the site of Parrsboro.

Parrsboro is a rockhounder's paradise. Sedimentary rock has preserved the wanderings of dinosaurs; volcanic basalt rock has created beautiful amethyst and quartz. Stilbite, jasper, and over 200 different kinds of agate, are found in the Bay of Fundy region.

From earliest times, native peoples used the Parrsboro Gap on their journeys from the marshes around Amherst. They would canoe up the River Hebert or walk the Boer's Back, proceeding through Newville, the lake system, and Farrells River until they reached the sea. They crossed the Minas Basin and travelled up either the Avon or Shubenacadie rivers, and then traversed more lakes until they reached the rocky Atlantic coast. To the Indians, Parrsboro was *Owokun* or *Awokum*, the "crossing-over point." Many of the stories of their god, Glooscap *(Good friend)*, took place here.

The first official Europeans to view this area arrived in 1607. This group was led by Champlain, who recorded in his log that they found an old, moss-covered iron cross just up the harbour from the island, at a place now called Crane Point. Champlain had no idea who these earlier Christians were. Perhaps the cross was left by Sinclair the Earl of Orkney's 1398 legendary expedition, perhaps by Basque fishermen. Champlain also found good-quality amethyst, which in those days almost littered the beach.

The legend of Maiden's Cove is set near Partridge Island. As the story goes, an Italian pirate named Deno, captured a ship and killed the crew. The only one spared was the captain's beautiful daughter. The ship was driven by a great storm into the Bay of Fundy and they found themselves at Black Point. A treasure in amethyst was gathered and the girl was put in a cave and closed in with stones. The ship sailed away. Later, Indians heard the girl's moans and fled. After awhile they returned to investigate. Pulling away the stones, they found her skeleton. The French heard the story and the legend was born. It is said that at certain times, one can still hear the moans.

As Acadian settlements spread in the early 1700s, the French used the Indian routes and began to operate a ferry service across the bay to the area of Grand Pre. The service was started by Jean Bourg and Francis Arsenau in the 1730s, who, at the time of the Expulsion in 1755, were imprisoned in Windsor's Fort Edward. They remained there until the turmoil ended.

During this time, a small group of British soldiers were marching through the gap to Fort Cumberland, unaware of the fate awaiting them, for beneath the Lakelands area at Gilbert Lake lies Ambush Maze—a swarm of tunnels left by subterranean glacial streams. When they neared the lake, they came to a natural meadow where Hanna Brook flows. Then came the war cries. From nowhere, it seemed, came Mi'kmaq warriors, first from the rear, and then from in front. The soldiers retreated up the brook, knowing it was a dead end. The hills that border the brook rise sharply to a height of 700 feet; the meadow itself is only a bit over 100 feet above sea level. The soldiers fell back to the shelter of the trees, up the canyon, and were trapped at a pool of water at the base of the falls. There they died. The warriors melted back into the tunnels and woods from which they had come.

Farmers in this area are aware of Ambush Maze. Marion Kyle speaks of a horse that broke through the ground over the maze. A ladder put through the opening went down about fifteen feet. Upon investigating, Kyle saw the maze of round tunnels spreading out in all directions. When the highway was paved in 1938, the highway department crashed into one of the tunnels.

The ferry service was re-established by Abaijah Scott after the French and Indian Wars. Its former proprietors, Bourg and Arseneau, worked for Scott. Many New England planters had arrived in the Annapolis Valley to settle vacated Acadian lands, and they helped provide the market needed to support the service.

In 1776, the American Revolution broke out. Would Nova Scotia, the northernmost part of New England, go with George Washington or remain loyal to King George III? The majority of the British in Cumberland were Yorkshire English; the rest were mainly New Englanders. Jonathan Eddy, a local New Englander from the Amherst area, dreamed of revolt. He travelled south to meet Washington but got little help. Nevertheless, he returned, raised a force of about 1,000, and besieged Fort Cumberland that autumn. It was important to break the land route to Halifax, so Partridge Island was attacked. A privateer full of men captured the tiny place and the ferry. That ended Scott's business. At Fort Cumberland, a relief force attacked Eddy at daylight and the rebellion was crushed. At Parrsboro, His Majesty's ship the *Vulture* re-established Imperial control. In June 1777, the *Vulture*, joined by the *Hope* and the *Mermaid*, transported troops to the Saint John River, where Eddy's force was defeated and destroyed. His base of operations, Machias, Maine was levelled by Sir John Collier's squadron, led by the flagship *Rainbow*.

On April 16, 1778, Parrsboro got its first permanent settlers, Mary and Silas Crane. They were soon joined by James Noble Shannon, formerly of Machias, Maine and Cornwallis, Nova Scotia. Shannon arrived at Partridge Island to open a store. He brought with him a youth apprentice: James Ratchford, who would, in time, become the "King Seaman" of the area and the husband of the Cranes' daughter, Mary.

Meanwhile, Partridge Island was attacked again. A Machias privateer tried to loot the store and village in June 1780. The eight men who landed were fought on the beach; two were killed and the rest captured by Lieutenant Wheaton's small defending force. The story has come down that the enemy captain was half Indian. Similar scenes were taking place all over the Maritimes, and the population became more and more anti-American. Privateers from Nova Scotia carried the fight to the south.

The defeat of the British in the American Revolution saw thousands of United Empire Loyalists arrive on the Nova Scotian coast—so many, in fact, that three new colonies appeared: New Brunswick, Prince Edward Island, and Cape Breton. Loyalists became a major group on the Parrsborough Shore. In 1784, the area became part of Kings County; by 1822, the township held 1,287 people.

PARRSBORO VILLAGE AT PARTRIDGE ISLAND, c.1818

The blockhouse surrounded by a palisade is on the right; in the background is the island.

In 1784, Governor Parr arrived for a visit; two years later, when the township was set up, it was named Parrsborough. The village was called Parrsboro'; the apostrophe was dropped when it was incorporated in 1889.

One of the Loyalists was Caleb Lewis, who was from Connecticut, but had moved his family to Vermont at the time the war broke out. His wife was killed, and he and his son Jesse became prisoners. Caleb escaped and made his way north. He was among the first settlers in Halfway River, where he lived alone. When his son and two daughters, Martha and Lois, arrived in Sackville to enquire about him, they heard that there was a Lewis near Partridge Island. Jesse crossed to Amherst Point, went over the Boer's Back, and showed up at the homestead one day, asking Caleb for work. His father, who did not recognize him, said he could stay a week but that there was no work for him. During that week, the youth did various jobs at Partridge Island settlement. At the end of the week, he took Caleb by surprise, saying: "Father, I think it's time I told you who I am." Thus Caleb was reunited with his son and two daughters.

A love story from these times survives. Twenty-five-year-old Ebenezer Bishop of Greenwich, Kings County, had seen eighteen-year-old Anna Lewis of Halfway River only a few times, but his mind was made up to marry the young woman. It was winter and the ferry was not running. He told his parents, Timothy and

Mercy Bishop, that he was going to see his friend Nathaniel (Nath) Loomer at Scotts Bay. It was night when he dismounted from his horse at Nath's door. His friend tried to dissuade him, but Ebenezer was determined. At daybreak the tide was slack, the ice pack motionless. The pair headed to the North Mountain and slid down into Amethyst Cove. Parrsboro was so close, just five or six miles away.

Ebenezer decided to aim for Cape Sharp. Nath had prepared a notched board or pole for him to use in crossing the ice floes. He had to be quick, before the pause in the tide ended. He set out, but about halfway across, the ice cake he was on began to sink. He threw himself flat on his face, using the notched board to haul himself to safety. The board also made a short bridge when needed. Finally, Nath saw a tiny figure silhouetted against the snow on the far shore, ascending the cliff. Ebenezer arrived at Jesse Lewis's farm near dusk. Anna, one of Jesse's seven children, was preparing supper.

"Will you marry me?" he asked.

"Of course, Ebenezer," she replied.

Three days later, he left on horseback, the animal on loan from Jesse Lewis. It was a long way up to Truro and back down to Greenwich, but he made it. In October, he returned for Anna and they sailed to her new home. When their baby boy was born they named him John Leander, after the Greek hero who had swum the Hellespont in ancient times. Ebenezer and Anna had six children and were married for 40 years. John Leander graduated from Acadia in 1843, became a doctor, and served in the American Civil War.

> The Nova Scotia Hellespont
> Four miles across from shore to shore
> the Minas Channel mauls
> The Fundy tides in narrow space between
> its rocky walls…
> Folk standing on the Parrsboro shore…
> raised at last a mighty cheer to greet
> him on the beach
> When questioned how he kept so cool,
> one answer would suffice:
> He'd say he got it from his dad,
> who braved the Minas ice.
>
> by Dr. Watson Kirkconnell
> (of Ebenezer 1784-1846)

The American problem persisted. A new war broke out; an offshoot of the Napoleonic Wars, it was called the War of 1812. Partridge Island had a large two-storey blockhouse surrounded by a palisade on the high hill overlooking the settlement. At the beginning of hostilities, a schooner was sent by the navy to re-ordnance the fort, and two naval brass cannon were left there. These would disappear in the 1930s.

The man who really built the Partridge Island settlement was James Ratchford, the nephew of James Noble Shannon. He was made Shannon's partner, and upon his uncle's death, Ratchford became the sole owner of the property. He made sure his children married well, thus expanding family influence. Ratchford's store became the focus of the basin and the Parrsborough Shore. His ferry, though it had competitors, carried the mail, and was a vital link to the county seat at Kentville. He built ships, kept a hotel, was Justice of the Peace and Colonel of the Militia. In 1812 he built the Brig *Parrsboro* at Partridge Island and sent it to the West Indies. It returned in 1813 with rum and molasses. The rum he sold to the militia that had been called up.

Politically, Ratchford was a strong conservative. He was against the annexation of the shore by Cumberland and discouraged building at Mill Village. Events were against him, however. Joe Howe's Liberals moved the post office to the present town site. In 1826, J. de Wolfe started a store in the same area. On May 4, 1836, Ratchford died and was laid to rest in the Anglican cemetery. He was 73 years old. Four years later, the township split, most going to Cumberland, and the Five Islands section to Colchester. Ratchford's house was purchased by Sir Charles Tupper around 1855 as a summer home, and has borne the name Ottawa House ever since. Tupper built a large ferry wharf and did his best to stem the abandonment of the site. In 1884, the home was purchased by the Cumberland Railway and Coal Company—12 years before Tupper became prime minister of Canada. As the years went by, Partridge Island declined. Today it is almost unoccupied, although Ottawa House Museum stands there.

Mill Village got its name from a tidal grist mill built by the Davison family on what is now Lower Victoria Street. From all over the Basin, people came with their grain and corn. The mill even had a kiln to dry oats, and manufactured oatmeal. More and more homes were built, businesses started, and churches erected. With the advent of more roads and stage coaches, the ferry route across the basin began to lose its importance. Even in 1803, a road of sorts had been built from Amherst to Truro by way of the Wentworth Valley. No longer was Parrsboro the hub of the province's three original roads: one through the valley; one from Windsor to Halifax; and one the trail from Fort Cumberland over the Boer's Back to Parrsboro.

On October 4, 1869, Parrsboro was hit by the Saxby Gale—so named after a prediction the previous year by Captain Saxby, R.N., that a severe storm would be caused by the moon's close orbit and planetary alignments. The warning was picked up from the London paper and was widely known, if not believed. When the tide surged into Mill Village, ships and docks were wrecked. It is claimed that a sand bar was created by the storm, connecting Partridge Island to the mainland at last.

Mill Village continued to grow. In 1877, the railway arrived from Springhill, and Parrsboro became a major shipping centre for coal and lumber. The shipyards were busy: Parrsboro had arrived. In 1901, the population reached its zenith with 2,705 people, although the death rate was still high, mostly due to disease.

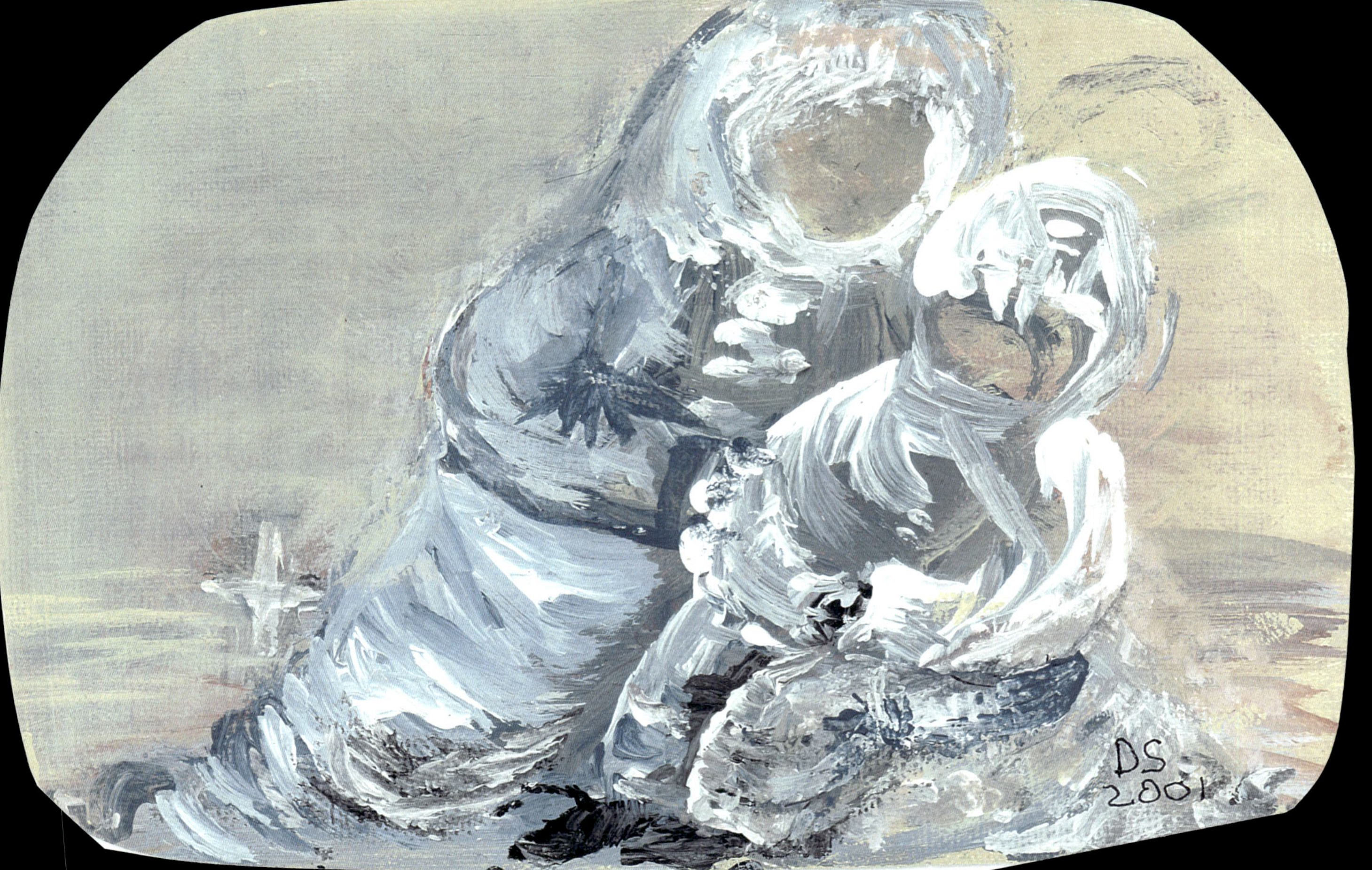
DS
2001

On Monday, September 12, 1881, thirteen-year-old Arthur Lee Dickinson, only son of John and Eliza, died. His illness began with depression and a headache. Soon, opening his mouth and swallowing became difficult. His muscles tightened and convulsed; spasms made it hard to breathe. There was no cure. Arthur Lee Dickinson died of tetanus. Nine days earlier, Arthur had been playing on the railway tracks not far from his Spring Street home in Parrsboro, and injured his foot. The bacteria entered his bloodstream and his fate was sealed. A photograph of the Dickinsons during the family's three-year stay in New Jersey, USA, shows Arthur dressed in his best Victorian clothes.

The family was not through with tragedy. On July 26, 1883, the couple lost Alice, their eleven-year-old daughter. She and two playmates had rummaged through a trunk of clothes belonging to someone who had died of diphtheria—another bacterial disease that swept both sides of the Atlantic in the late 1880s. The common method of contagion was through coughing. Alice contracted diphtheria, as did her two friends, and all three died shortly thereafter.

The Dickinsons' last child, Maude, grew to womanhood and became a Parrsboro teacher and an accomplished painter. She married Walter Fullerton on October 9, 1907, and moved out west, where she died in childbirth at the age of 41. Both she and the infant she bore, Arthur Burgess, perished on the same day in 1919 and were brought back to Parrsboro for burial.

ARTHUR LEE DICKINSON, 1868–1881

His father was a sixth cousin of Abraham Lincoln. The night of the assassination, Lincoln was watching the play *Our American Cousin* at the Ford Theatre in Washington, D.C. One of its lead characters had been developed by an American who had run a theater in Halifax, Nova Scotia.

THE DICKINSONS

Maude (Dickinson) Fullerton: 1877–1909 with her parents Eliza and John Dickinson

**ELIZA (?) AND
ALICE
DICKINSON
(1872–1883)**

Bacterial diseases we no longer worry about today robbed many children of their lives. The fear of losing a child would have been constant and families that lost more than one child were not unusual. The Southampton cemetery holds a marker for the children that Moses L. and Matilda Harrison lost in February, 1877, to diphtheria:

> Harold A. died Feb. 3, 1877 aged one year, nine months
> Mary E. died Feb. 6, 1877 aged three years, six months
> Herbert R. died Feb. 18, 1877 aged two years, five months
> Arthur L. died Feb. 23, 1877 aged six years, three months

Below the marker appears the despondent prayer: *"Meet us there."*

The *Chignecto Post & Borderer* of Sackville on May 5, 1888, referred to Parrsboro as the "Black Hole of Nova Scotia" because of the town's large number of whiskey shops. Another story carried in 1885 dealt with some young boys who were vandalizing the area, breaking windows. They were finally caught, but the case could not be proven so they were released. Finally, they were caught for a break and enter. It was discovered that one of the boys was only ten years old and that they wanted money for liquor. This discovery was grist for the temperance mill.

During World War One, Nova Scotia joined all the other provinces, save Quebec, in restricting alcohol to medical and scientific use. Quebec joined this group in 1919, with the exception of wine and beer. Ontario permitted local wine. In 1920 Nova Scotians, by plebiscite, approved prohibition. In the United States, the Volstead Act came into force that same year. Many people still wanted to drink, however. Moonshine was made everywhere, or so it seemed. Maritime vessels picked up liquor in St. Pierre et Miquelon and carried it to the American coast. Families made fortunes in rum-running, and so did the underworld. Prohibition ended in Nova Scotia in 1930. Only Prince Edward Island held out longer—until 1948.

The twentieth century was not kind to the town. The days of wooden ships were gone, the coal mines in Springhill closed, and the rail line ceased to operate in June 1958. Occasional fires burned down sections of the main street, the museum at the base of blockhouse hill was bulldozed, and the population declined. For a time there was a large sawmill in Riverside run by Scotts, but it, too, closed. As the century neared its end, there were the amazing dinosaur finds and the establishment of the town as a tourist centre.

PARTRIDGE ISLAND, 1888 "Awaiting the Tide." The ferry *Hiawatha* is waiting, and so are the passengers. They are at Tupper's Snag by the island.

HIAWATHA   The Partridge Island Pier around 1900 with the *Hiawatha* ferry at anchor.

PRINCE ALBERT   The *Prince Albert* ferry, named for the consort of Queen Victoria, c.1900. In 1917, it caught fire at the pier but was saved.

THE M.V. KIPAWO

The M.V. *Kipawo*, the last ferry on the Parrsboro-Valley run. It operated from April 1926 to January 1, 1941, at which time it ceased operations and became part of the Canadian Navy operating in the British colony of Newfoundland. One of its jobs was tending submarine nets. At the end of the war, it was erroneously declared unseaworthy and remained in Newfoundland until 1982, when it was returned to Parrsboro and grounded near the bottom of Whitehall Hill to become a theatre. The word Kipawo stands for Kingsport (Kings), Parrsboro (Cumberland) and Wolfville (Kings).

THE ENTRANCE TO PARRSBORO HARBOUR

The lighthouse at the entrance to Parrsboro harbour, c.1900. Today an automated lighthouse is used.

PARRSBORO, AS SEEN FROM WHITEHALL, c.1900

The large building by the bridge was the sail loft, where sails were made. In the upper right are the Parrsboro schools.

COAL WHARF A view of the coal wharf looking down the harbour, about 1900. The Bryson tug is on the left. Beside it is the so-called miners' wharf, by the lower track. The coal-burning train is on the upper track. Between the tracks is an inlet where barges were built.

PARRSBORO The lower track with empty coal cars and several vessels at anchor.

VIEW OF PARRSBORO The main track from Parrsboro proper, heading toward the area called Whitehall and the coal dock. In the background a vessel is seen loading timber.

THE BRIDGE TO RIVERSIDE The bridge to Riverside around 1900, before the building of the aboiteau. In those times, there were a few shipyards further up the river on the left. When vessels needed to pass, the bridge would swing open. The last ship to go through was the *St. Anthony*. It had been launched the previous night and squeaked through at noon to enter the main harbour. Not long after this, the mechanism to control the bridge was in such bad condition that it became a fixed structure and no longer opened.

THE HARBOUR The river at Parrsboro with the tide in, near the Riverside bridge, c.1900. The track forked here: the left went to the lumber pier, the right to Whitehall and the coal wharf.

THE HANDLEY-PAGE *ATLANTIC* CRASHES

The crash of the Handley-Page at Parrsboro, July 5, 1919.

**THE
HANDLEY-
PAGE
ATLANTIC**

The Handley-Page *Atlantic*, which crashed in Parrsboro on July 5, 1919, and left again on October 9. Powered by four Rolls-Royce engines, it had been in Newfoundland to attempt the first air crossing of the Atlantic. It had to wait for a part, however, and competitors Alcock and Brown won the fame. The plane was returning to New York when it ran into difficulties. The people turned their lights on and made bonfires along the harbour to warn of danger while the plane circled overhead. All thirteen cars in the town assembled at the racetrack; at daybreak the plane landed but overran the track, blew a tire, and landed on its nose. Eventually it was repaired and able to leave. It was a large plane: 62 feet long, with a wingspan of 126 feet.

THE *ATLANTIC* The Handley-Page gets ready to leave Parrsboro in October, 1919.
REPAIRED

THE BAND-STAND, PARRSBORO The pride of Parrsboro for many years was its Citizens' Band, which had a bandstand erected in the 1890s. The town got electricity at the end of 1897. and by the end of the next year 90 homes had electricity. The photo was taken around 1900.

ABOVE THE HARBOUR

Lower Main Street, Parrsboro, c.1925, looking south. The impressive new post office opened in 1913; just beyond it is the World War Monument. The age of automobiles had arrived.

BEACHED WHALE AT OTTAWA HOUSE

In front of the Ottawa House (the old Ratchford building) near Partridge Island, c.1901. The event: a beached whale. The dead whale was towed out to sea by the tug *Susie* and dynamited.

SAILING PAST PARRSBORO'S LIGHTHOUSE IN THE EARLY TWENTIETH CENTURY

PARRSBORO: WEST BAY

THE COAL WHARF

The Whitehall section of Parrsboro as seen from the air in 1931. On the right is the coal wharf with a vessel getting coal from the chute. Just above the wharf are three buildings: on the left was the office, called the Company House; on the right was the Pump House, which supplied water to ships. Behind these are the railway tracks. The trains proceeded above and to the left of the wharf, and then came down to unhook the coal cars. When empty, the car would have its brake released and gravity took it down the length of the dock. A track to the right (not shown) was used by engines backing up to pick up the empty cars. Loaded cars waited at a siding, visible just above the Company House. In the harbour are several lumber scows and a three-masted vessel. On the lower right was the repair shipyard—the so-called "ship hospital." One of these vessels is the *Bentley*. The tops of several other ships can be seen in the inlets.

RIVERSIDE WHARF, C.1900

The old stone munitions building survived from the early nineteenth century until the mid-1960s when the town of Parrsboro had it destroyed. At one time it was a school, then it became a museum. Of the town's decision to bulldoze the building over the bank, exhibits and all, Mrs. Lottie Wheaton said the following:

> I went to the town hall when I first heard what they were planning to do. They said they would notify me when the bulldozer was to be used. They didn't call me. If they had I would have stood in front of that machine. We owned that building at one time. I wanted the people to see the past. Stolen! When I went out after the bulldozer had been there I cried. They had pushed it right over the bank. I went down into the rubble and brought out this nineteenth-century cloth—all muddy! It was so stupid!

I have often thought of these lines from Joseph Howe, a famous nineteenth-century Nova Scotian:

> A wise nation preserves its records…gathers up its monuments…decorates the tombs of its illustrious dead…repairs its great public structures and fosters national pride and love of country.

THE ROAD TO WEST BAY   A late Victorian view of the ammunition building, Partridge Island. Both it and, at another time, the blockhouse foundation were bulldozed.

UPPER MAIN STREET   Looking toward the upper part of Main Street, Parrsboro, around 1900. The spire of St. George's Church is visible on the right.

WEST BAY, PARRSBORO, C.1900 Here, lumber boats loaded from scows from Parrsboro and the surrounding area.

ON THE BEACH The Victorians of Partridge Island: gathering on the beach.

A BOY, DOUGLAS CUTTIN, WITH HIS DOG.

GOING FOR A DRIVE

AN OLD MAN BY HIS HOME

MRS. CUTTIN AT THE CHURN, PARTRIDGE ISLAND, C.1900

OTTAWA HOUSE, PARTRIDGE ISLAND SETTLEMENT, C.1900

STORM SURGE: TUPPER WHARF DAMAGED

BUSY PORT   Sailing ships by Ottawa House, with Cape Blomidon in the background, c.1900. Partridge Island is to the middle right.

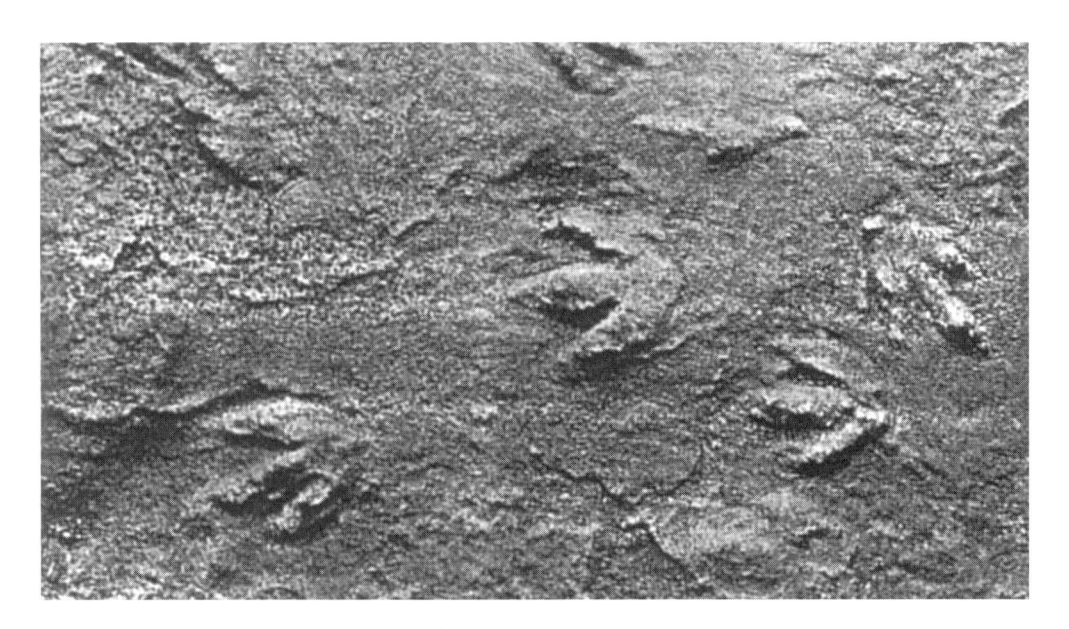

WORLD'S SMALLEST DINOSAUR TRACKS

The world's smallest dinosaur tracks were found in Parrsboro by Eldon George at Wasson's Bluff on March 31, 1984. George had been up to Moose River collecting fossils. As he drove his four-wheel buggy back, he pulled in at an outcrop of rock to escape the cold. He saw a small track (at left in the photograph) and began to investigate. These footprints were of a 220-million-year-old animal called *Coelophysi*, and are the smallest dinosaur prints ever found. The marks were from a juvenile, which would have been eight inches long and six inches high. Fully grown, this dinosaur would have been between six and eight feet (1.8 and 2.4 metres) long.

That year, George also found in the Jurassic sandstone a giant crocodile footprint, that of the *Otozoum* species, a terrestrial animal with large legs. One footprint was 21.5 inches (55 cm) long and 15 inches (38 cm) wide. This creature would have been about 34 feet (11 metres) long and 12 to 15 tons. This find was also at Wasson's Bluff. Among George's other amazing finds was the first freshwater horseshoe crab found in Canada, dating from 300,000,000 B.C. On August 2, 1989, Dr. Paul Olson, of Columbia University, said of this find:

> The horseshoe crab, about the size of a child's fingernail, was a missing link in the area's natural history. In Nova Scotia we have fossils which span over four hundred million years…and there are particularly good deposits in the Carboniferous, Triassic and Jurassic….But over the years very few fresh water invertebrates have been found…first of its kind in Nova Scotia although researchers have been scouring the beaches for more than one hundred and fifty years. The horseshoe crab shows tremendous similarity to similar finds in Europe; and really is quite a precious…discovery….These….probably were never common along the area, so the find is important.

Near his home, in 1962, George made another major find: a 300-million-year-old *haplolepid* fish fauna, the first and the oldest of its type found in Canada.

There have been numerous spin-offs from Eldon George's work. Parrsboro now boasts the ultra-modern Fundy Geological Museum and a yearly "Rock Hound Roundup" that have made the town famous. Much of the credit for this success is due to Eldon George.

ELDON GEORGE, 2001 A model of *Coelophysis*, the world's smallest dinosaur tracks, found by Eldon George.

Chapter 5

God's Jewel Box

**ADVOCATE HARBOUR
TO THE PARRSBOROUGH SHORE**

ROAD BUILDING

Road Building, c.1900, at Fox River, just below the Fox Point Road. No engines—just horses, wagons, shovels and man power.

Advocate Harbour to the Parrsborough Shore

On the southwest coast of the county is "little Cape Breton"—a rugged, ocean-washed coastline where high hills reach a sea coast hugged by small villages. At Advocate Harbour, a large harbour with dyked shores encloses 450 acres. The villages lie at the edge of the reclaimed land, where two small brooks meander. To the east is Cape d'Or, to the west Cape Chignecto, home of Nova Scotia's largest provincial park. It covers about 10,000 acres (4,200 hectares) and rises to a height of about 600 feet (185 metres). Spectacular trails navigate the cape.

As the glaciers retreated, exposing the present land, it became a stopover for nomadic Mi'kmaq, who thought of it as the god Glooscap's herb garden. Nearby, Spencers Island was Glooscap's overturned kettle; Isle Haute was formed when his dogs chased a moose into the bay. As the moose drowned, it begged for help. In compassion, Glooscap turned it to stone, making it eternal. Over the years, evidence of Indian occupation has been found—everything from arrowheads to the surprising discovery in 1898 of a tiny birch bark coffin containing the remains of a child that was brought to light when a grave was being dug in the cemetery.

Off the north coast of Scotland lie the Orkney Islands. According to legend, Henry Sinclair, Earl of Orkney, sailed his fleet into Nova Scotian waters in 1398. He seems to have gotten on well with the Indians, and it is said they brought him to Advocate Harbour, where he wintered. During that time, Sinclair's crews constructed a ship. Six hundred years later, a small model of this Scottish vessel, on loan from Massachusetts, was displayed in the county. In the 1500s, the Portuguese explored the Atlantic Coast. On their maps, the words *Rio fondo* (deep river) denote today's Bay of Fundy.

In 1607, Champlain arrived. It is believed that he gave the place its name, *le avocat* (Advocate), for Marc Lescarbot, the colony's lawyer. Nearby he found copper; hence the name Cape d'Or. Several decades later, the French settled in the area, building dykes and small settlements that thrived until the Expulsion of 1755. One tragic story of that time gave rise to the name Refugee Cove.

On the shores of the North Mountain, forming one flank of the Annapolis Valley, is a small village called Morden. Many of the villagers of Belleisle, fleeing from the New Englanders and the British, came to this area. Their leader, an old man named Pierre Melancon, was known as a loner and a woodsman who had crossed the bay before. Their plan was to escape across the bay to Advocate Harbour. Several children had died by the time they reached Morden (also called French Cross). Here, in the winter of 1755/56, they became stranded; their ammunition gone, their only available food was mussels. As many as 200 perished. In March, an old Indian and a youth arrived. On St. Patrick's day, Melancon, ill, and the young Indian crossed the bay for help. Immediately the French moved to their rescue. On the bow of the relief boat was Melancon, but when they reached French Cross he was dead. His wife died shortly afterward. The survivors that arrived at Refugee Cove settled there and built a look-off, now called French Lookoff. The Acadians at Advocate Harbour, however, were driven out.

Around 1765, Lady Harley of Scotland sent some Scottish settlers to Advocate Harbour, but supplies failed and so did the settlement. It seems their vessel was wrecked on the opposite shore from Advocate Harbour at Scott's Bay. The people landed there and were helped by a local hunter. They then faded into history. In 1780, Robert Spicer of England settled here, and was soon followed by those who were defeated in the American Revolution: the United Empire Loyalists.

The 1800s were good to the area. In 1840, the Parrsboro township had 2,113 people and was detached from Kings County, most of it joining Cumberland. This political move reflected the development of roads going toward the north. Nevertheless, the sea, with its fishing and shipbuilding industries, was of extreme importance. All along the shore were shipyards: Diligent River, Port Greville, Spencers Island, and Advocate Harbour. Between 1870 and 1920, these shipyards produced 115 three- and four-masted schooners—one quarter of all that were built in the province. The most famous vessel constructed here was undoubtedly the brigantine *Amazon* of Spencers Island, better known to history as the *Mary Celeste*. This was the first large ship to be built at Spencers Island. The builder was Joshua Dewis; timber was supplied by the Spicers.

Dewis was born in 1815 in Economy, the same year that the wars against Napoleon I finally ended. He was known as a hard worker who had been making boats since his teens. As an adult, he moved to Spencers Island, drawn by its timber resources. Construction of this two-masted ship, began in 1860, and it was completed in May 1861. Its home port was given as Parrsboro; its length was just under 100 feet. It was a good time to have a ship, as there was money was to be made because of the American Civil War.

The ship's first voyage had bad luck. Captain Robert McLellan took it to Five Islands, where timber was loaded for London. They went down the Bay of Fundy but the captain took sick. At the New Brunswick port of Quaco, a doctor came aboard, and recommended they return to Spencers Island, which they did. McLellan died there. The new captain, John "Jack" Parker of Walton (Hants County), took the ship to London and then on to Spain. Unfortunately, in the English Channel, they struck an English brig, sinking it. No lives were lost. For the next two years, their route was the Atlantic, from the Caribbean to the Mediterranean. Upon returning to Nova Scotia, the *Amazon* got a new skipper, named Thompson. Three years later, the ship had its last captain. His name was Murphy. In November 1867, he sailed to Cape Breton to pick up coal. November is often a stormy time in these waters, and the ship was driven ashore in a storm at Cow Bay, Cape Breton Island. After that, it was purchased by American interests and renamed the *Mary Celeste*. It was now three feet longer.

The Mary Celeste

The *Mary Celeste*, when it was known as the *Amazon*. Captain Parker had this picture done when he was at Marseilles, France at the end of 1861. The name *Amazon* can be clearly seen streaming from the top of the mainmast, with the Union Jack from the foremast. The other flag is the red jack of the British Merchant marine. Seven years later, the *Amazon* became the American *Mary Celeste*. The original drawing was given to Fort Beausejour and a copy given to Conrad Byers of Parrsboro by the family.

The mystery of the ship began on December 4, 1872, when it was found under sail and abandoned in the Atlantic between the Azores and Gibraltar. It had sailed with ten people on board: Captain Ben Briggs, a Bible-reading temperance personality; his wife Sarah; their two-year-old daughter, Sophia; and an experienced crew of seven. They had left Arthur, their young boy, with his grandmother so he could go to school. Their cargo was 1,700 barrels of medicinal alcohol. The *Dei Gratia*, which came upon it, salvaged it. They noticed two hatch covers over the cargo open, and that the beds were unmade. The log book was present, but the chronometer, sextant, and some papers were gone.

They brought the ship into Gibraltar and a hearing was held, but the mystery was not solved. Most now believe that the captain, unfamiliar with an alcohol cargo, had smelled fumes and had the vessel abandoned by row boat, attached to the ship by rope. The wind came up, storms came in, the rope broke, and all ten perished. As for the ship, it reverted to its American owners and in 1885, it appears, was deliberately wrecked off the coast of Haiti for insurance money. The wreck was discovered in 2001. There is some confusion about the ship's name.

Marie Celeste is the name of a vessel in a fictional tale by Sir Arthur Conan Doyle of Sherlock Holmes fame. Written in 1885, the story was based on the *Mary Celeste* story.

When the Saxby Gale hit on October 4, 1869, the village of Advocate Harbour was taken by surprise. The dykes were in poor condition, and when they burst, water poured into houses. The next year, the dyke was rebuilt.

HORSESHOE COVE Cape D'or, Horseshoe Cove, former shipping point for the Colonial Copper Company, c.1918. The vessel is the *Cape d'Or*, run down by a steamer off LaHave, Lunenburg County, in 1925. Four crew members and the captain's wife died in that accident.

CAPE D'OR The row houses lie abandoned in 1915–16.

In 1897, Advocate Harbour was all abuzz. The Colonial Copper Company was building a complex at Cape d'Or to mine copper. There was an ice-free anchorage at Horseshoe Cove, and several seams of copper. A representative of the company stated:

> Hanway Lode is situated…east of Bennett Brook Lode. I blasted samples across it and found it to be 200 feet wide, and fully 70 feet of this carries copper in paying quantities…the formation is visible to the eye for a distance of 6,500 feet north. It is dipping vertically and also extends under the channel…contains copper equal to 44 pounds per ton of 2,000 pounds of ore. Bennett Brook lode is 55 feet wide and the formation is visible to the eye for 6,500 feet north…exposed immense bluffs, some being 250 feet above high water mark. The top part is sedementary red sandstone. The lower part, trap rock [and] conglomerate between [these last two] copper in paying quantities.

Three main shafts were sunk to a depth of 200 to 300 feet. Houses were built, and a short rail line was put in. A concentrating plant, shipping facilities, offices, and boarding houses were constructed. The mine was abandoned in 1905. As time passed, the copper veins became mixed with rock, and instead of making $6,240 a day after expenses, the company was losing money. All the houses were taken to other villages, except the manager's home, which eventually burned down. Today everything is gone. But a magnificent walk exists down the road to the light house at the cape. The twentieth century saw the entire coastline in a period of decline, decade by decade.

ADVOCATE HARBOUR

The first bridge here was built in 1835 and replaced in 1904. It connected the east end with central Advocate Harbour. The picture shows high tide as viewed from the main village.

ADVOCATE HARBOUR, c.1910

The two-span iron bridge was erected in 1904. Just to the right of the vessel are two stores; the larger is the Atkinson store. It drew on a population of 1,598 (1901 census) in the polling district that included Spencers Island but not Apple River. On the far right is the school, built in 1905. It had 181 students and only four teachers. The largest class had 54 students!

T.K. BENTLEY

Named for its owner, the *T.K. Bentley* was built in 1920 in West Advocate Harbour.

**WALTER CALLOW
1894-1958**

Walter Callow was born in Parrsboro in 1894. During World War One he was in an airplane crash in Ontario and suffered a severe back injury, from which he never recovered. His military career was over. Callow moved to Advocate Harbour, where as a businessman he ran a hotel and a lumber business. In 1931, both his wife and his mother died. He was left with a ten-year-old daughter. In 1937 he was admitted to Camp Hill medical hospital in Halifax. He never left. He had become a quadriplegic. Several years later, he went blind. By 1952, his legs had become so deformed they had to be amputated. Callow did not give up. Supported by volunteers, he began a wheelchair bus line for shut-ins. He gave a new freedom to those confined to their homes by disability. He died in 1958. "I believe," he said, "that after a man dies one question will be asked [by God]: What did he do for others?" I am sure that God is well pleased with what Walter Callow accomplished while on this earth.

In August, 2001 a plaque was unveiled where Walter Callow lies in the Advocate Harbour cemetery. One of the Callow buses was parked nearby, and the president of the Walter Callow Veterans and Invalids League, William MacDonald, spoke. Overhead there was a military fly-pass; on the ground a bag pipe played. Stan Spicer told of visiting Callow and seeing a verse he had printed over his bed: "Don't swear. It's not that I give a damn, but it sounds like hell to strangers." For those who knew him, he is remembered for his sense of humour as well as his accomplishments. For those who did not, he is remembered nation-wide for his "indomitable soul" and his undying faith, two qualities reflected in these words:

> Sunset and evening star,
> And one clear call for me!
> And may there be no moaning of the bar,
> When I put out to sea,…
> The flood may bear me far,
> I hope to see my Pilot face to face
> When I have crossed the bar.
>
> —Alfred Lord Tennyson

July 1910, LaPlanche Street, Amherst

The event is the start of the Fifteen Mile Race at Amherst's Old Home Week. The runners are from left to right: Sterling of Saint John, G.B. De Mar of Massachusetts, Home of Massachusetts, Hackett of Massachusetts and Advocate Harbour's Fred Cameron. Cameron will win it. Hackett was second, De Mer third, Sterling fourth, and Home fifth. The event took place not long after Cameron won the fourteenth Boston Marathon on April 19, 1910. Local papers were jubilant. His time: two hours, twenty eight minutes and fifty two seconds. This was a full minute ahead of the second place finisher. One hundred and eighty-five had started the race.

"I saw the others were not running at a pace that was fast enough for me. Down where I come from we have sand roads and as there is a lot of snow there I could not train for this race as long as I would have liked to, so I really had only two weeks to get ready. I was too heavy, weighing about 120 pounds. I lost seven pounds during the race."

The mayor of Boston, John FitzGerald (grandfather of J. F. K.), had given him the keys to the city and hosted a dinner for him. Amherst claimed him.

Cameron was born in Advocate Harbour on November 11, 1886. His father, Joshua A. Cameron, was a prosperous farmer. As a youth Cameron liked to sprint over the countryside. In 1907 he appeared in Amherst at the Labour Day sports event and came in second in his first races. Next Labour Day he was in Springhill where he won two of his three races. In 1909 Tom Trenholm became his trainer, and a year later he was in Boston.

After that victory he raced professionally but had given it up by the time World War One started. He married an Amherst woman. They moved and finally ended up in Vancouver, British Columbia where Fred Cameron died in 1953. They had no children.

FRED CAMERON—WINNER OF THE BOSTON MARATHON

SPENCERS ISLAND

Spencers Island around 1920, with the four-master *Susan Cameron* at anchor. The ship was built in 1919 at New Glasgow, Pictou County. The old shipyard wharf is visible on the lower right. The *Mary Celeste* was the first of more than 30 vessels built here. The largest was the *Glooscap*—more than twice as long as the *Mary Celeste*. Built in 1891, the *Glooscap* was eventually cut down and converted into a plaster rock barge. It sank in New York harbour in 1924. The anchorage dock was destroyed in a gale in the 1960s.

THE RUPERT K.

The *Rupert K.* under construction at Spencers Island in 1920. It burned at Campbellton, Restigouche County, New Brunswick, in July 1921.

**ISLE HAUTE
LIGHTHOUSE**   Isle Haute Lighthouse, which operated from 1864 to 1956, when it burned. There is now an automated light there. The island was formed from a lava flow during the Jurassic period—about 200 million years ago. About 5,000 years ago, at the same time that the pyramids were being built in Egypt, rising water and shifting land seperated the island from Cape d'Or.

ISLE HAUTE

One interesting spot on the Advocate Harbour coastline is Isle Haute ("High Island"), as it was named by Samuel de Champlain around 1604. This island, with an area of less than a square mile, lies about six miles from land. It rises sharply from the sea, reaching a height of well over 300 feet. At the landward end, at sea level, is a large spit of beach covered in driftwood. The beach contains a small lake bordered by an area called Indian Flat. Arrowheads have been found there, perhaps from the Indian dog feasts said to have been held there. The only anchorage for the island is located off the beach, where there used to be a wharf.

A lighthouse was built in 1874. The lighthouse keeper and his family lived year round on the island. The first to man it was Captain Card (to 1889), followed by Judson Reid (to 1904), Percy Morris (to 1941), his son-in-law Donald Morris (to 1946), and finally John Fullerton (to 1956).

In 1890, 17-year-old Mildred Balsor was hired to teach the four children on the island. Her classroom and her bedroom were both located upstairs in the lighthouse building. For leisure she explored the island with its wildflowers and birds. One night in late October, her father decided that she would not spend the winter there. As Mildred made her bed, she spotted a white sail and knew it was her father coming to take her back. Sadly, she obeyed him. The trip home was miserable; the wind had died down and oars had to be used. The men covered their blistered hands with jute bags and eventually made it to Port Greville, and from there home.

Ella (Morris) Fraser recalls her experiences on the island as a child in the early part of the twentieth century. She and her siblings had most of their schooling in Advocate Harbour. They boarded out, but some years their mother stayed in the village as well. Ella was always glad to get back to the island. Ella's father raised about 60 sheep on the land he'd cleared as a small farm. Dulse, berries, and fish were also available. The only wild animals were rabbits so tame they would come right up to the doors. Many people came to visit—300 on one July 12, Orangeman's Day.

Communication with Advocate Harbour was done by a bonfire system: one bonfire meant no problems, two meant illness, three meant send a doctor, and four, a death. There were two deaths: the first was a helper, the second was Fred Harding. On his twentieth birthday, Harding climbed down a cliff to look for some gull's eggs and fell fifty feet. Morris and another man searched for him all night. At daybreak, after launching a small boat, they found his body. There was always talk of Captain Kidd's treasure on Isle Haute. In 1952, eight Spanish doubloons and a skeleton were found by Edward Snow, an American historian.

A New Brunswick company introduced fox ranching—which killed off the rabbits. The foxes died out or went over the ice to the mainland. The mystique this island holds still draws tourists like a magnet, as they flock to the new Chignecto Park.

ISLE HAUTE AND CAPE CHIGNECTO, 1777

Isle Haute and Cape Chignecto from Joseph DesBarres's *Atlantic Neptune*, 1777, a magnificent collection of charts, plans, and views assembled by this Minudie landlord while he was in London, England.

**JOHN
CHIPMAN
KERR
1887–1963**

Down the road and around the corner from the Road Building was the home of John Chipman Kerr (1887–1963), one of 94 Canadians who has won the military's highest award—the Victoria Cross—since its inception in 1856. Born in Fox River on the Parrsborough Shore, one of the six children of Robert and Elizabeth Kerr, John, with his brother Roland, headed west in 1906. The brothers ended up in Alberta in the Peace River country at Spirit River. Both enlisted in the army in 1915, serving in the 49th Edmonton Regiment. Their fates would be different. Roland was killed, while private John "Chip" Kerr was awarded the Victoria Cross for his heroic efforts on September 16, 1916, at the Second Battle of the Somme in France. With a finger blown off, Kerr, acting as bayonet man, raced along the trench shooting, chasing the enemy into the dugouts. Believing that they were surrounded, 62 unwounded Germans surrendered. Ignoring his wound, Kerr continued until he and two others turned over the prisoners. Back in Canada, he married an Englishwoman and they raised their family on Canada's West Coast. John Kerr died in British Columbia. A mountain in the Victoria Cross Range is named after him.

**PORT
GREVILLE**

Port Greville, showing several vessels. A century ago Port Greville was busy with plankers, caulkers, sawyers, blacksmiths, yard hangers, and horses. There were four independent shipyards in the area: the Cochrane, the Elderkin, the Graham, and the Wagstaff and Hatfield. Eighty-four known sailing vessels were constructed here: 76 schooners, 6 brigantines, 1 bark, and 1 barkentine. The yards also built over 50 fishing ships, several tugs, numerous barges, and 12 minesweepers that were used during World War Two.

EATONVILLE Eatonville was located in the north of the Cape Chignecto peninsula, along the Eatonville Brook. The dock, mill, store, and shipyard were located at the harbour. Upstream was a dam to provide a holding pond used to help drive logs down the stream. The Eatonville mill, pictured here c.1900, was purchased in 1897 by the Elderkins of Apple River. The capacity of the mill was 60,000 feet of deal per day, but they preferred to cut less, waste nothing, and strive for quality. The forest of Cape Chignecto tended to be damp from the fogs and salt breeze, which greatly reduced the danger of fire. Their store, built in 1899, was an impressive structure with hardwood floors, mosaic squares, two large plate-glass windows, a jewelry display case, and even a telephone. The head clerk and bookkeeper, C.D. Dunfield, had been a private in General Kitchener's army from 1896 to 1898 when the rebellion in Sudan was suppressed by the British. Today, Eatonville is abandoned.

**B.R. TOWER
LAUNCHING**

In 1901 Diligent River and its satellite communities had a population of 915. The first settlers arrived in about 1777. Two were military officers. When Governor Parr visited in 1784, he found the settlers so industrious and diligent that he named the place Diligent River. The first child born in the region was Lieutnant Taylor's son, John Parr Taylor. The governor had said that if they named the boy after him, he'd give them another thousand acres.

The photograph shows the *B.R. Tower* being launched around 1920. It was wrecked in the West Indies shortly afterward. Earlier the *L.C. Tower* had been launched, but it, too, was short-lived, destroyed by a German submarine near the Irish coast during its maiden voyage on May 14, 1915. The cook on the *L.C. Tower* loved his fiddle. When the Germans boarded, the crew were all ordered to leave. The cook asked if he could go below to get the instrument, but the answer was "No." The Germans placed two bombs and some tins of gasoline on the ship and sank it. Until his dying day, the cook lamented his lost violin. There was no loss of life, however, as the crew were permitted to row ashore.

**CUMBERLAND
QUEEN
LAUNCHING**

The launch of the 634-ton *Cumberland Queen*, built by Charles Robinson, at Diligent River in 1919. The tugboat is likely the *Mildred* from Parrsboro with Captain Harteny Wasson. On the *Cumberland Queen's* first voyage, the ship went from Parrsboro to Swansea, Wales, with lumber, and from there took coal to Italy, then salt from Spain to Massachusetts, then it returned to Parrsboro.

**APPLE RIVER
LIGHTHOUSE**

The 1901 census recorded 609 people in the Apple River-New Salem area. In 1991, Apple River had a population of 78, New Salem about the same. The name Apple River is said to have come about when a vessel laden with apples was wrecked and the fruit was strewn along the water's edge. The sawmill was important in these villages. Apple River was also a ship-building centre. Thirteen vessels were constructed there: the first was the *Victory*, built in 1845, a small boat of ten tons; the last ship was built in 1909. Pictured is the Apple River Lighthouse. It was built in 1870 and burned in the late twentieth century.

**WARDS
BROOK,
C.1900**

Bringing the logs down Wards Brook, just west of Port Greville.

**THE SHULIE
LIGHTHOUSE**

**MOOSE
RIVER**

Moose River, a small village to the east of Parrsboro, around 1920. A flume has been built below the dam. A trough of lumber was constructed in a V shape, then water was put through it to freeze into ice. The men are waiting for the cut trees to be sent sliding down this ice luge—a dangerous job.

WARDS BROOK

This house started life at Cape d'Or. It was moved to Wards Brook and later to the valley area, both times by water. A few years ago it was moved again, this time by land. Moving houses by barge was a simple solution to the problem of "moving house." In 1921, Jake Goldstein, a merchant of Advocate Harbour, decided the future was better in Joggins, so he loaded his ten-room house on a barge and brought it around Cape Chignecto and up the bay. It was a calm June day; horses moved the house to the scow, and Mrs. Goldstein stayed aboard baking bread while chickens perched on the verandah. The scow left Advocate Harbour at six o'clock that morning and arrived in Joggins around noon. Crowds at Joggins watched the tugboat—and the smoke from the Goldstein's chimney. At Lower Cove, they waited for low tide to ground the scow, then winch and horses were used to bring the house to lower Main Street, Joggins. It was placed on its new cement foundation where the home still stands. Not even an ornament was broken during the trip.

SAND RIVER

Between Joggins and the Advocate Harbour region is a stretch of road running about 40 kilometres through the woods. Two villages once thrived here by the sea: Shulie (a corruption of the French word *joli*, "pretty") and the smaller Sand River. Both were lumber centres. Today, Sand River, pictured above, is abandoned. Shulie has a permanent population of two. In the 1930s a sluice system operated to take cut lumber from Cochrane's inland mill to the sea for shipment. Basil Pettigrew's mill here was running two shifts during the same period.

THE *EUGENE OWEN MACKAY*

The *Eugene Owen Mackay*, earlier sister to the *Cumberland Queen*, as it is being built at Diligent River, c.1918 by Pugsley and Robinson. Sold to Reinhardt, *Eugene Owen Mackay* was engaged in the usual trades until caught in a gale north of Bermuda on January 4, 1926, where it was abandoned as it sank.

Shulie

Shulie, in the days when it prospered c.1900. The settlement was established around 1850. The founders were Joseph Reid, Job Seaman, and James Patterson and his brothers. The Pattersons opened a sawmill. By the turn of the twentieth century, Shulie boasted a church, a school, a dock, and many homes, as well as the mill. A lighthouse was constructed in 1904. Mrs. Leander Canning recounted how the news of the *Titanic* reached them from Joggins when a peddlar came through yelling "Did you hear? Did you hear? The Tic Tac sank." At the time of the Halifax Explosion, two Leander girls went to the School for the Deaf in Halifax. When their father heard what had happened in Halifax, he snowshoed to Joggins in a raging snowstorm and caught the train, getting as far as Bedford. From there he snowshoed into the city. He was the first father to get to the school. His daughters had been kneeling for morning prayers when the explosion hit on that morning of December 6, 1917. They were saved from much of the flying glass, though one suffered a cut on the head. Their father took them out of the city and back to Shulie, where they spent the rest of the winter.

Shulie was abandoned during the 1920s. A huge fire swept the Shulie area around 1921. Next, the mill closed. The lighthouse was shut down in 1924. Around 1933, Harry Pettigrew of Southampton took his mill to Shulie. His son, Earl, said "The houses looked like the people just walked out and left them, taking only their furniture. The road was really a trail that you could get a car through. One house was used as the cook house, another as a bunkhouse. The mill sat on a hill and lumber was shot down to the harbour by way of a sluice." Shortly after the Pettigrews left, the houses were torn down.

By 1976, the road was so bad that the section from Shulie to Joggins was taken off the government tourist highway map, although it was eventually put back on. Bill Casey, the local Member of Parliament, realized that for the area to make progress, a new infrastructure had to be in place, so the Shulie Road was paved. Now Advocate Harbour and Joggins are well connected.

Chapter 6

The Town That Would Not Die

SPRINGHILL

EARLY NINETEENTH-CENTURY SPRINGHILL

Edwardian Springhill as viewed as one goes up Monument Hill, with the post office on the right. This building was lost in the fire of 1975.

Springhill

Springhill is located near the geographic centre of Cumberland County and at the far west of the region's coal fields. It sits atop a hill that rises to a height of 651 feet (198 metres). There were 14 official mines in the area. When the mines closed after the disasters of 1956 and 1958, many felt that the town would die. It didn't. The people remained united and determined. Small industries, a museum, and a prison were built; and for a time even a small coal mine operated.

The first mention of coal being sold from Springhill occurs in 1834. Lodewick Hunter, who got his land grant in 1827, was operating a small mine located on the banks of a small brook on one of the large seams of coal. In that year, 16-year-old Charles Dixon Lockhart was living in Westbrook and working as an apprentice to Thomas Leak, carriage maker. He was sent on a 30-mile (50 km) round-trip with a cart to pick up some Lodewick coal, which Leak needed for his forge. In 1849, Lockhart again went to Springhill and saw the Albion Mining Company exploration team inspecting the area. Among its members was William Patrick of Maccan.

Patrick was back again with two Americans in 1863, but the trip came to naught. The interior remoteness of the area, its rugged landscape, and the dim prospects for a railway were likely discouraging. Patrick had already opened the Victoria Mine (1858) in River Hebert, and a Maccan mine (1861). In 1863, he discovered coal, leading to the development of the Chignecto and St. George mines.

By 1871, the area consisted of only five small houses and perhaps 20 people. The old folk in Southampton used to say that Springhillers had to come to Southampton to vote. But all that was to change. The Spring Hill Mining Company came into being by an act of the provincial legislature on April 18, 1870. Soon coal was being produced but had to be carted through the woods to Athol or Salt Springs. Much of the coal was being used in the construction of the Intercolonial Railway. In the fall of 1873, the first trainload of coal was sent from Springhill Junction; one year later, the main line was opened. Between 1873 and 1877, a railway was built to Parrsboro to transport coal. The first shipment went out on March 15, 1877, and the boom was on. By 1881, the population reached 800; by 1891, it was 4,813. The town had an ugly setback, however, on February 21, 1891.

Mother Coo

Ellen Creighton was born in 1833 in Northumberland County, New Brunswick. In 1854, she married Edward Coo. She moved to Pictou County, where she became well known. Some said she was a witch, others felt she was a harmless old woman who made a living gazing into a tea cup. Her notoriety increased when she was said to have predicted a disaster that occurred at Westville's Drummondville

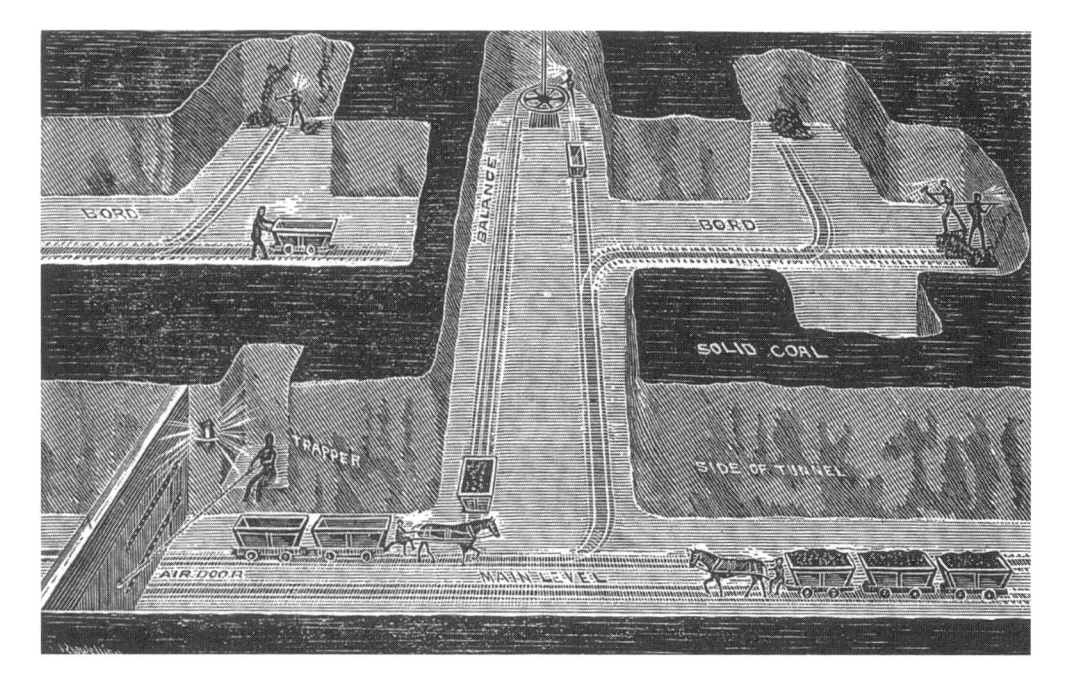

MINE INTERIOR   The interior of a late nineteenth-century coal mine showing the trapper boy, the horses and track, level, balance, and bords.

Mine in 1873. That mine caught fire and was being emptied when it exploded. Two hours later, a second explosion sealed its fate. That disaster took 55 lives. Miners spoke about Mother Coo as they worked: some called her a silly old hag, but her reputation continued to grow. Then she spoke of an upcoming explosion at Stelleraton's Ford Pit. Some said she'd given the hour and day; some said she should be thrown in jail. But many worried, and on that doomsday some women succeeded in getting their husbands to stay home. The Ford Pit exploded on November 12, 1880. Fifty-two died in the subsequent inferno. Mother Coo became famous throughout the province.

Therefore, it was no surprise that the Cumberland Railway and Coal Company became a bit nervous when they heard that Mother Coo had predicted a fatal explosion in the Springhill Mines before Easter 1891. The mines were checked by a workmen's committee of experienced coal miners. They reported that there was no cause for alarm, but still people wondered. Henry Swift was the newly appointed underground manager for the collieries. He never expressed fears of a gas explosion to Conway, the underground manager of Mine No. 1, but he was worried about Mother Coo's prediction. On Saturday, February 21, 1891, he was ready to leave the mine for the day when he said to a miner: "I've checked it all out. I can't find anything wrong, but I'm going to check again." They found him buried under rubble; he was the last one to be brought out.

Conway left the mine that morning at 10:30, having visited what was to be the explosion site. There was no gas. Manager Swift had told him Friday night that he was to attend a meeting in Maccan. Conway had gotten as far as Springhill Junction when he got the news: "The mine had exploded and men and boys were all in it." Mother Coo died in Everett, Massachusetts in 1912.

THE EXPLOSION OF 1891

At the base of the Spring Hill on the Southampton side were the three main mines: No. 1 (East Slope), with 300 workers; No. 2 (West Slope), with another 300; and No. 3 (North Slope), with a workforce of 400. Unfortunately for the miners in No. 2, there was a tunnel connecting the two mines at the 1,300-foot level. An additional 350 men worked on the surface.

There were two other mines as well. The Syndicate (where the miners' museum is today) was begun in 1886, but closed because of a fire two years later. The other was the Aberdeen at the foot of Junction Road (sometimes called No. 4). It opened in 1888, but was inactive in 1891. Because of a lack of empty coal cars, No. 3 was not working on that day. There were 600 people in the No. 1 and No. 2. At noon, they stopped for a half-hour dinner break, then went back to work. Above ground, a cold rain fell. At 12:43, the explosion came.

At the 1,300-foot level on the No. 1's west side, Thomas Wilson fired a shot blast into the wall of the bord (the working shaft) to loosen coal. He hit methane gas. The gas and the coal dust exploded, shooting flames through the mine and, by way of the tunnel, into Mine No. 2, where several died. The main cause of death in Mine No. 2 was the poisonous gas. About 50 died in the other mine because of the after-damp gas, also poisonous. The rest were killed by the actual

THE NO. 1 MINE The engine house is on the left, the carpenters' shop, which was used as a morgue in the 1891 explosion, is on the right. The flag flies at half-mast.

explosion and its related damage. Manager Swift was at the pit bottom where timber was being lowered. When he heard the sound, he rose to his feet and started to make his way out. He got about ten feet when he went down, perhaps overcome with after-damp, and was buried by falling rock. Seventeen boys, aged 16 or younger, died that noon hour. One who survived, James Liddle, said this about the disaster:

> At the time of the explosion I was at the bottom of number one slope, and all at once my light went out, and then I smelt gas, and knowing something was wrong, I at once tried to make my escape, and almost got up to the surface when I was struck on the back by a boom [roof rafter]. When brought to the bankhead I was senseless. When I came to, my legs, from body downwards were paralyzed. I was taken immediately to my boarding house, where I stayed until removed to the hospital [Victoria General, Halifax], where I have been slowly improving. Hope to get around again in a few months. The things published about me are wrong. They have got my name David instead of James and my age 22 instead of 16 years, which it ought to be. This is about all the information I can give.

Underground manager John Conway must have been frantic. His 13-year-old son, William, was working in the mine, driving his horse-drawn rake. The horse, badly burned, fell on him, and this saved him from the flames. The rescue party at the 2,000-foot level thought all the living had been removed when they heard a feeble cry: "Mother! Mother!" They turned the horse over, and found the boy uninjured and only slightly affected by after-damp.

The removal of the dead was a pitiful scene. After the explosion, two brothers, David and James McVey, had started to run but were overtaken by after-damp. They were found locked together in each other's arms so tightly that they were taken to the surface that way.

> Killed in the bowels of the earth
> Where none could hear their cries
> Or listen to a last request
> Nor close their dying eyes

From "Springhill Mine Disaster 1891," originally sung by Ruth Metcalfe.

Several miners who escaped said they never heard such cries as those of some of the men affected by after-damp, who realized that they must lie down and die. Alexander J. McKay, a young man from Tatamagouche, had only been working in the mines for three days when he died. Families waiting at the surface saw their hopes shattered as no one was brought up alive after two o'clock. Instead, there were 121 dead and 4 so badly wounded they died later. There was no miracle. One hundred and sixty-eight dependent children were left behind. The mines were rebuilt and reopened.

QUEEN VICTORIA

Queen Victoria made a personal financial donation to Springhill after the 1891 explosion.

The Explosion of 1956

On Thursday, November 1, 1956, at 5:07 P.M., the afternoon shift at the No. 4 Mine had been at work only two hours. The 20-year-old mine had been having trouble with gas build-ups. Six cars being hoisted to the surface became unhooked and crashed down the steep slope, hitting the steel transmission cable and churning up coal dust. Near the 4,400-foot level, the cable was crushed. The resulting blast shook the town; flames belched from the bankhead, throwing timbers high into the air. The sound was heard by hunters up to fifteen miles away. At the bankhead, five died and four immediately burned, two died later. The other two, Norman Boss and Alan Skidmore, were blown through a door and onto the ground. Those who died at the bankhead were: Joseph Crummey, Ben McLellan, Pleaman Pyke, Lester Nelson, David Vance, William Jones, and Lester MacDonald. People poured toward the mine. One hundred and eighteen miners were underground, and 30 men were lost.

In a short time, rescue workers entered the mine to save those who were trapped. Two, William Ferguson and Alex Spence, died from the gas; Ferguson was down only 150 feet when he lost consciousness. Soon rescue crews from Pictou and Cape Breton counties arrived. The gas was so thick, it was feared that no one had survived. The air indicator on the surface, however, showed air was being taken from the line. The newspaper headline on Friday was grim: "Little Hope for 116 Trapped Miners." That was the message from Dosco General Manager Harold Gordon. Saturday's headline was worse: "In All Probability the Men are Dead."

Meanwhile, Deputy Overman Conrad Embree led a large group of men from deep in the mine. He had a rubber air hose cut so that each man could breathe. They also had plugged their noses to keep from inhaling the gas.

The Explosion of 1956

The ruins of the bankhead and the entrance to No. 4 Mine.

On the surface there was fear of a second explosion. Still, it was reported that tapping on a pipe had been answered by trapped men. Just before noon on Saturday, another group of miners led by Wilfred Brown finally made their way to the surface. Brown himself had spent a short time at Embree's shelter. Meanwhile barefaced miners and rescue workers, keeping their faces close to the ground, continued to penetrate the mine in spite of the gas, talking to each other, pinching each other to keep awake. For awhile the rescue was halted while repairs were made to the ventilation system.

At nine o'clock on Sunday night, it was confirmed that about 50 men were awaiting rescue at the 5,400-foot level. They had a door at each end, benches to sit on, and two barrels of water. They were afraid, at first, to open the door in case more fumes entered. Hot drinks and food were sent down to them. At about one o'clock on Monday morning, the first of the survivors reached the surface. Meanwhile, two more men were found at the 5,700-foot level. It was indeed a miracle. Even Philip Brown, the youth who was the pump-man at the mine's bottom, the 8,100-foot level, was saved. Gas tests showed no one else could be alive. There were fears of another explosion, so on Monday morning the mine was sealed. On January 18, 1957, it was re-opened to remove the twenty-six bodies. And then it was closed forever. The explosion of 1956 had killed 39 men. (See Appendix C for the names of those who died in 1956.)

THE BUMP OF 1958

On Thursday, October 23, 1958 (almost a year after the town was hit by a major fire), the stress in the tunneled underground maze of Mine No. 2 grew too great. A bump like an earthquake came at 8:05 P.M.: the floors rose, the ceilings of the mine squashed downward, the sides collapsed, and gas came in. As far away as Southampton, the ground re-adjusted. There were 174 men in the mines. By Friday morning, 72 had been brought up alive. The crowds at the pithead thinned as only a few more survivors were found. Early Tuesday came a news flash: a dozen more had been found. They were at the 13,000-foot level.

Hugh Guthro, 31, who had been trapped for three days in the 1956 explosion, was tossed up against the celing when the bump hit in 1958. He knew miners were dying near him. He remained conscious but was buried up to his chest in rubble. Others helped him out.

> At first we tried to dig our way out. Soon we got so weak we had to stop. We had no food and no water since Sunday or Monday. The last thing we ate were a few crusts of bread. We rationed our lights and cut some batteries off the dead men but on Monday we were in darkness. I'd say Caleb Rushton kept us going.

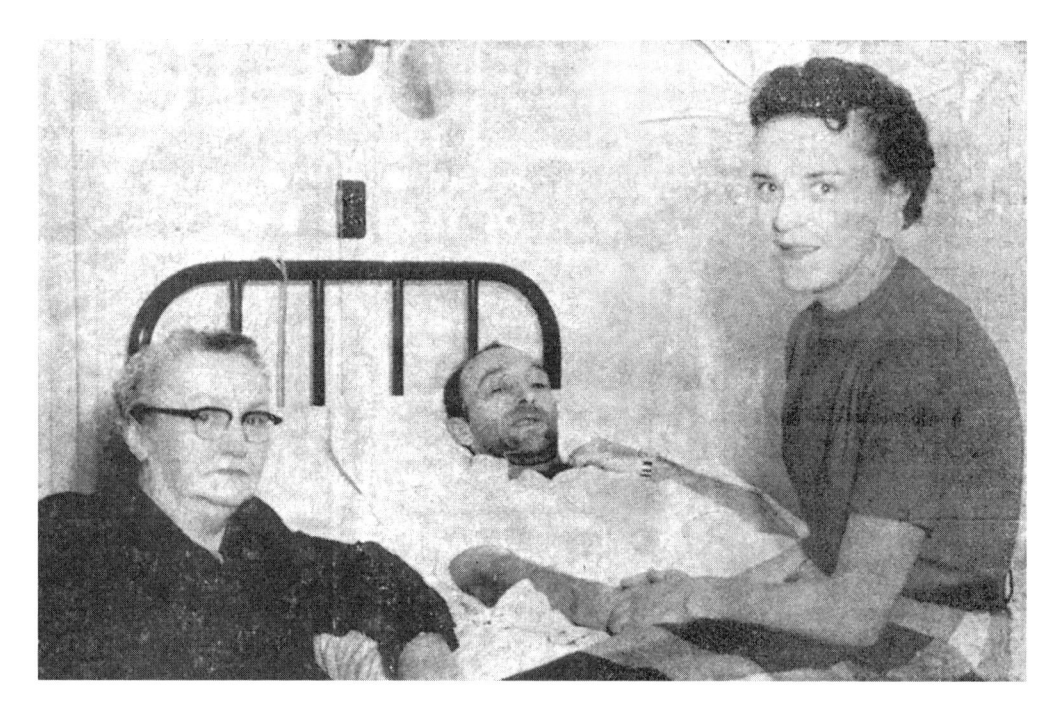

A SURVIVOR OF THE 1958 BUMP

Hugh Guthro, recovering from his six days underground, with his mother and his wife.

After four or five days he thought he could hear distant noises, but they seemed far away: "Since some of our boys had already given up hope, I didn't mention it," said Guthro. Rushton sang hymns. "I tried to lead them as best I could and sometimes we were all singing together. I guess our favourite was Stranger of Galilee. It seemed to help," Rushton would later say. There was no gas in their tomb, and fresh air seeped in. Joe McDonald said:

> I just lay there and I prayed. When the bump hit, I was blown clear up to the roof. I bounced right down again. Then the floor started buckling, and lifted me to the roof again. It sagged and dropped a foot or two and I lost consciousness. When I came to I was lying in a rock pile. Men were dying and groaning and moaning all around. I could hear them.

Rushton said two lunch pails were found. The men nibbled on sandwiches for two days. Bowman Maddison added: "I kept feeling over for the man right next to me to make sure I wasn't alone. I would know he was still with me and both of us were alive." Gorley Kempt said of his six days in the bowels of the earth: "I dug a cubbyhole in the rubble. I thought there might be a way out but there wasn't. And soon I knew there wasn't. And so I laid there day after day and I prayed. And I'm not a praying man."

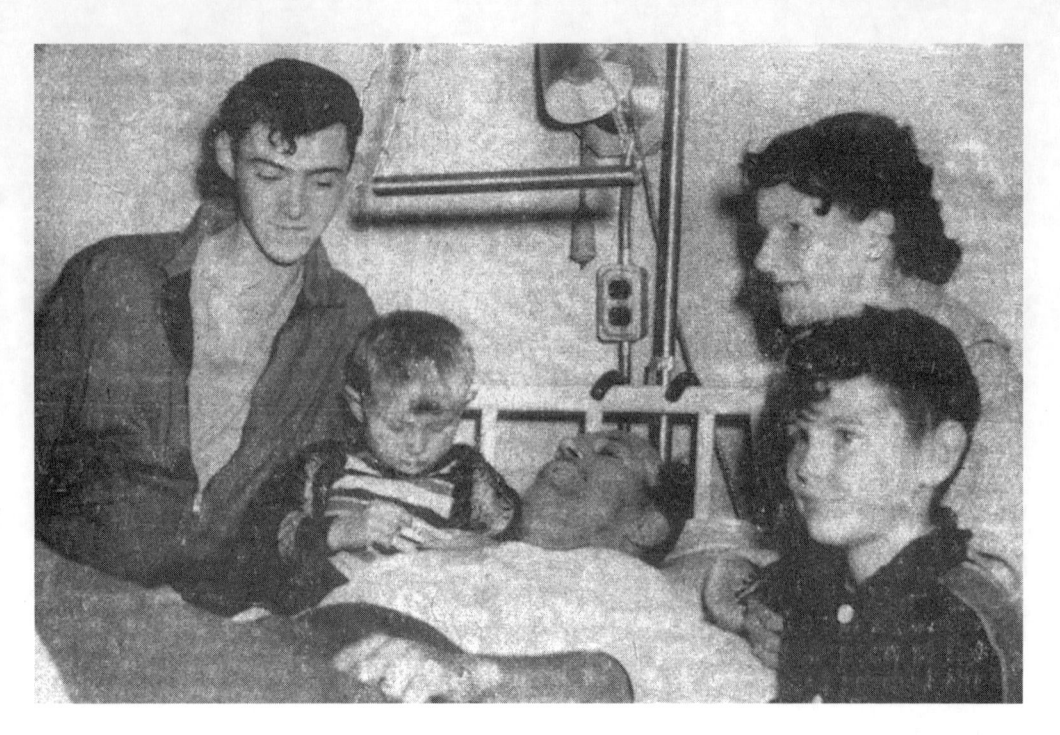

SPRINGHILL, 1958 Joseph MacDonald, who spent six days entombed in the mine, recovering from a broken thigh. With him are his wife and three sons: Gary, 16; Michael, 4; and Gordon, 11.

At midday on Wednesday, Kempt was lying with his mouth to the air tube pipe. which they used to tap signals, when he heard a click on the pipe. He shouted. Contact had been made.

Doctor Burden supervised the insertion of a 60-foot copper pipe through the air tube. Through it fresh water was pumped in. He told them to drink it slowly. This was followed by hot coffee and tomato soup. "We just sat during the last three days. We were all so happy to see them [the rescuers], the whole bunch of us just started to cry. I'm done with coal mining," said Guthro.

It had been six days. Gorley Kempt, Harold Brine, Joseph Holloway, Wilfred Hunter, Larry Leadbetter, Levi Milley, Theodore Michniak, Caleb Rushton, Bowman Maddison, Eldred Lowther, Joseph McDonald, and Hugh Guthro were free.

At 6:45 P.M. on Friday, Prince Philip arrived. Don Ferguson, mine survivor, his cheek still marked by the gash he'd received, would tell the prince near the mine site, "I'm glad to meet you, sir. Just seeing you has done me a world of good. I'm glad to see you here." At the pithead, Prince Philip insisted on entering the mine, and when he emerged the people cheered. He walked from tent to tent, talking with the Salvation Army and Red Cross workers. Earlier, he'd visited the miners in the hospital. Levi Milley's daughter asked for and got an autograph.

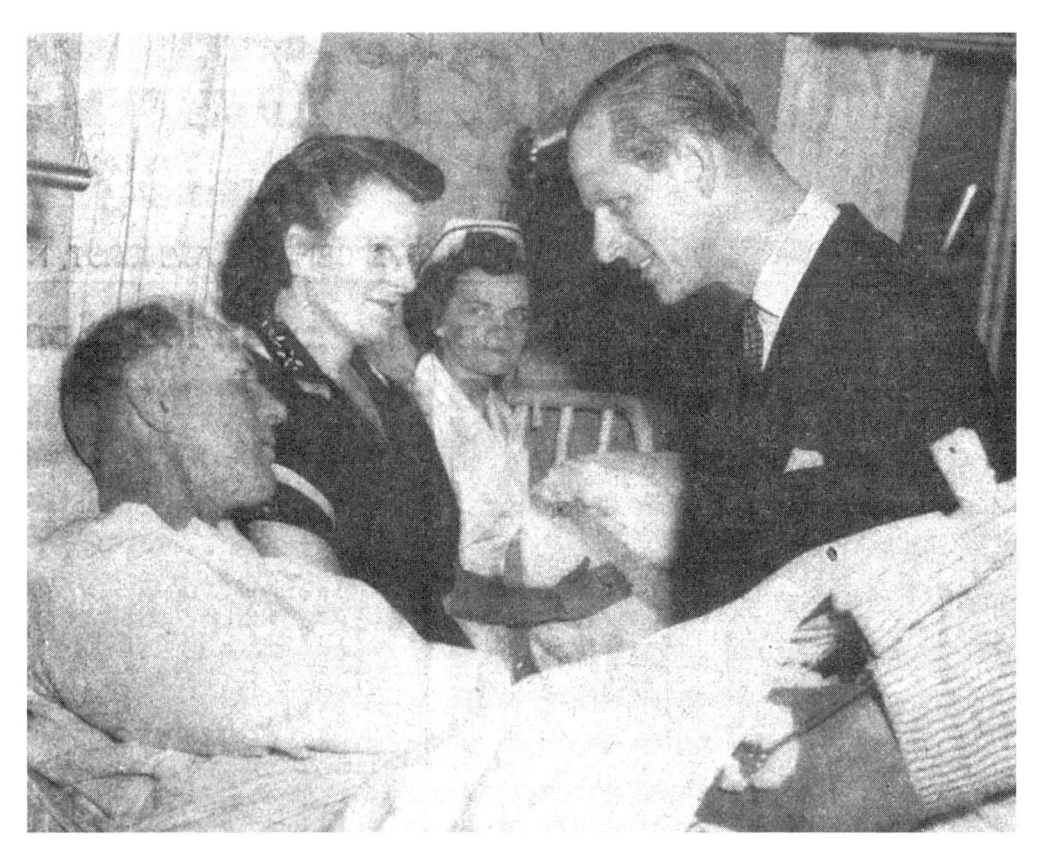

A ROYAL VISIT, 1958 Prince Philip speaks with Joe MacDonald, who survived the explosion of the No. 4 in 1956 and spent six days entombed in 1958.

Next, the prince visited the makeshift hospital at the armoury. He also made a surprise visit was to the home of Mrs. Harold Raper. Only a few hours earlier, she'd buried her husband who had been killed in the bump. Of this visit, she recalls:

> I was watching him on television and all of a sudden there he was in front of me. The prince just walked in and held my two hands in his until I got up. I didn't hear him come in the door. He was so nice. I told him there should be a padlock put on that mine down there. I told him all that we wanted was an industry to keep our men out of the mines. He had a cup of tea. He wasn't like a royal prince but like a prince of a man.

When Prince Philip left Springhill, the townspeople felt they were not alone. Underground, seven more men were still alive. They would be buried eight and a half days. They re-emerged on Saturday morning with their story. They were: Doug Jewkes, Maurice Ruddick, Herbert Pepperdine, Garnet Clarke, Frank Hunter, Byron Martin, Currie Smith. There had been an eighth man, Percy

Rector. His arm had been held fast by fallen rock and wood. He pleaded for them to amputate his arm, but they had no knife. He died on October 28, at the 13,000-foot level.

They were in the same general area where the other twelve men had been. At first the gas came, but it cleared up in time. Two days after the bump, they found Martin in a small hole. He was 100 feet (thirty metres) from the rest. Thinking it best not to drag him in the dark, they left him in his space. "God must have saved this little hole for me," Martin later said. Ruddick, "the Singing Miner," would tell of singing hymns and of drinking urine. Jewkes spoke of the sound of rats. "We were all lying down," said Ruddick, "when we heard John Calder hollering 'How many there?'" They were found. (See Appendix C for the names of those who died in Springhill Mine No. 2)

The mine never reopened. The old era was gone. The population fell as families moved away, but the town hung on. In 1975, another fire swept Main Street, but they rebuilt. Small factories, a prison, the miners' museum, the Anne Murray Centre (named for the famed Springhill singer), and a community college helped the town survive. It would not die.

THE ANNE MURRAY CENTRE, SPRINGHILL

SPRINGHILL An aerial view taken August 27, 1931, showing the bankheads of Mine No. 2 and No. 4 looking toward Southampton. The photograph was taken by Richard McCully and David Reid.

ALL SAINTS HOSPITAL AT THE END OF THE NINETEENTH CENTURY

THE SPRINGHILL POST OFFICE IN THE LATE NINETEENTH CENTURY The post office of Springhill, which stood on Main Street about a third of the way up Monument Hill.

SPRINGHILL RACETRACK IN THE LATE NINETEENTH CENTURY

The town on the hill as viewed from the racetrack in front of the main mines. On the lower right is St. Andrews Church.

THE MINES

The bankhead for No. 2 and No. 4 coal mines around the 1940s. In 1891, many died in No. 2 from gas that entered the mine when an explosion rocked the connecting No. 1 mine. In 1958, an underground bump destroyed No. 2. No. 4 was destroyed by the explosion of 1956.

THE MINES   No. 2 and No. 4 mines: a general view.

THE STRIKE   A twenty-two-month coal mine strike hit Springhill from 1909 to 1911. This postcard picture is dated June 12, 1910. Canadian forces are seen marching down Monument Hill through the business district. In the middle left is the post office tower. In the left background are the mines.

ALL SAINTS COTTAGE HOSPITAL, SPRINGHILL, 1906

When the 1891 explosion took place, there was no hospital in the town. W.C. Wilson, rector of All Saints Anglican Church, pushed for a hospital. It opened on November 1, 1893, with Wilson as the first superintendent. It was soon replaced by a larger 27-bed structure, shown in the photograph. A school of nursing was also part of the facility. In 1924, a wing was added. It was replaced again in 1963.

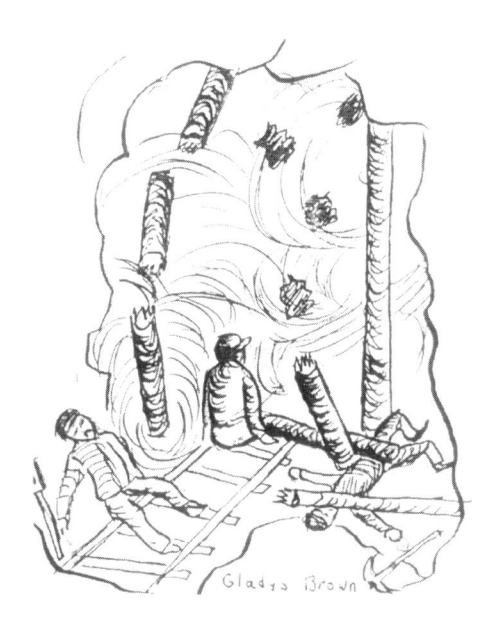

DISASTER, 1891

The 1891 explosion in Springhill, as shown in a drawing by Gladys Brown.

THE TOWN THAT WOULD NOT DIE

THE MINE COMPLEX

The Springhill collieries of mines No. 1, 2, and 3, c.1900.

THE NUMBER OF DEATHS IN THE SPRINGHILL MINES (1876–1969):

Number Two:	174
Number One:	139
Number Four:	66
Number Three:	34
Number Six:	6
Number Seven:	4
Syndicate:	4
Aberdeen:	1
surface (not specified):	8
unknown:	7
TOTAL:	443

THE
SPRINGHILL
MINERS'
MONUMENT

"The White Miner" was unveiled on September 11, 1894, by Governor General Lord and Lady Aberdeen, and the Right Honourable John Thompson, prime minister of Canada. Thompson went on to London to be knighted by Queen Victoria, and died an hour later.

A ROYAL TOUR, 1939 King George VI and his wife, Queen Elizabeth, getting ready to board the train that will take them through Springhill. The year is 1939.

A VILLAGE MILL, c.1900 Leamington, the village to the south of Springhill, c.1900. The John E. Gilroy Shingle Mill on the south branch of the Maccan River was water-powered, then was converted to steam by Herb Stonehouse. Neatness wasn't always a virtue!

Epilogue

There is a Future Here

Writing this book has been a labour of love, for this area is my home. I have seen it struggle to adapt to the massive changes of the second half of the twentieth century. And what a struggle it has been—mines are gone, shipbuilding has ceased, and forestry and fishery resources have been severely depleted. The population figures in Appendix B show the effects of this struggle. Farmers are paid little for their crops, with the exception of the blueberry industry; Cumberland County produces perhaps half of the world's wild blueberries.

The area has done its best to change the mental map of outsiders who, as tourists, see the sum of Nova Scotia as Halifax and the Cabot Trail. In southern Cumberland, the scenery is superb: there is a vast network of roads and villages, fine historic displays, and nature trails. Parrsboro boasts the Fundy Geological Museum with its dinosaur display as well as the historic Ottawa House; Port Greville has its beautiful Age of Sail shipbuilding complex; Advocate Harbour its Cape Chignecto Park; Joggins has the Fossil Centre; River Hebert its heritage models and miners' museums. I have a passion for Minudie, with its King Seaman Museum and all the trails to explore. Springhill has its miners' complex, where one can go underground, and the Anne Murray Centre. Beaches are everywhere, and they change throughout the day with some of the world's highest tides.

Though often forgotten by Halifax and Ottawa, this area has been noted by our sovereigns, from Queen Victoria's aid money for the 1891 explosion in Springhill to Prince Philip's spontaneous visit in 1958 when the bump in the mine devastated that town. Many Nova Scotians know little of this area and the promise it had and still holds. There is a future here.

Appendix A

A List of Deaths in the Mines by the Sea
Joggins, River Hebert, Maccan, Chignecto
compiled from various sources

	Name	Mine	Date
1.	Robert Palmer	Marsh	January 29. 1869
2.	Henry Hickman	Marsh	January 29, 1869
3.	Charles Lockhart	Scotia	December 5, 1874
4.	John Hudson	Chignecto	February 17, 1883
5.	W. Patton	Chignecto	February 17, 1883
6.	I. Burrows	Chignecto	February 17, 1883
7.	Charles Burke	Joggins	July 11, 1883
8.	Dan Lockhart	Chignecto	October 30, 1883
9.	Albert Fraser	Black Diamond	September 16, 1890
10.	Amos Brown	Joggins	September 8, 1892
11.	Jabez Pike	Joggins	September 13, 1894
12.	Emile LeFavour	Joggins	February 17, 1898
13.	William Long	Jubilee	October 11, 1902
14.	Alex Liddle	Chignecto	February 3, 1904
15.	John Livingston	Marsh	April 16, 1904
16.	J. W. Fraser	Marsh	March 13, 1905
17.	Ira Ripley	Joggins	February 13, 1906
18.	Lemuel Boudreau	Strathcona	February 5, 1907
19.	John Cormier	Chignecto	November 11, 1907
20.	John Coleman	Joggins Mine	December 24, 1908
21.	George Sawyer	Joggins Mine	December 24, 1908
22.	Charles Ackles	Jubilee	February 2, 1909
23.	James Carson	Chignecto	November 2, 1909
24.	Fidel Landry	Joggins Mine	May 16, 1910
25.	William Stevens	Chignecto	June 20, 1910
26.	Arthur Wood	Chignecto	June 20, 1910
27.	E. Hardcastle	Great Northern	May 12, 1911
28.	Sam Farrell	Scotia	March 9, 1912
29.	Sebastian Farrell	Scotia	March 9, 1912
30.	Daniel Rector	Kimberley	May 21, 1912
31.	Dan H. McNeil	Scotia	August 11, 1912
32.	John E. Stephenson	Scotia	September 26, 1912
33.	Philip Nicholson	Joggins	November 1, 1912
34.	William Hurley	Lawson	January 1, 1913
35.	John Burbine	Joggins	January 10, 1913
36.	Vincent Riclselts	Scotia	May 7, 1913
37.	James Porter	Black Diamond	April 16, 1914
38.	Thomas White	Scotia	December 16, 1914
39.	Thomas Jennings	Joggins	June 1, 1915
40.	Henry Gibson	Joggins	June 14, 1915
41.	Clifford Moffat	Minudie	November 25, 1915
42.	Wilfred Theriault	Scotia	July 13, 1916
43.	Alex Arseneau	Victoria	October 27, 1916
44.	John Doyle	Kimberley	December 4, 1917
45.	William Hall	Joggins	June 19, 1918

46.	John Martin	Jubilee No. 2	August 21, 1918
47.	Samuel Terris	Joggins Mine	December 2, 1918
48.	Jasper Wood	St. George	February 6, 1919
49.	William Main	St. George	March 25, 1919
50.	John Thurbid	Scotia	June 20, 1919
51.	Ernest Jowett	Victoria	1920
52.	Robert MacAloney	Maple Leaf	April 5, 1920
53.	Walter Purdy	St. George	April 8, 1921
54.	George Neal	Jubilee	May 12, 1921
55.	George Shannon	Joggins	November 2, 1921
56.	James Nicholson	Minudie No. 2	April 24, 1922
57.	Lawrence Lang	Minudie No. 2	April 26, 1922
58.	Edward Flemming	Jubilee	September 22, 1922
59.	Matthew Whalen	Jubilee	September 29, 1922
60.	William Cope	Victoria	November 7, 1922
61.	John McKeigan	Jubilee	December 6, 1922
62.	James O'Regan	Joggins	January 30, 1924
63.	David Edgar Wood	Strathcona	August 29, 1924
64.	Soloman Allen	Maple Leaf No. 4	October 31, 1928
65.	Philip E. Brine	Victoria, No. 2 slope	September 17, 1930
66.	William Burke	Victoria, No. 2 slope	September 17, 1930
67.	Simon Fowler	Victoria, No. 2 slope	September 17, 1930
68.	Emil Kralicek	Victoria, No. 2 slope	September 17, 1930
69.	Clarence McGrath	Victoria, No. 2 slope	September 17, 1930
70.	Wilfred White	Victoria, No. 2 slope	September 17, 1930
71.	William White	Victoria, No. 2 slope	September 17, 1930
72.	Thomas Jones	Victoria, No. 4 slope	May 11, 1931
73.	Adolph LeBlanc	Victoria, No. 4 slope	May 11, 1931
74.	Sanford Legere	Victoria, No. 4 slope	May 11. 1931
75.	George Quinn	Victoria, No. 4 slope	May 11, 1931
76.	Samuel Rector	Victoria, No. 4 slope	May 11, 1931
77.	Charles Stevens	Victoria, No. 4 slope	May 11, 1931
78.	Daniel J. Boudreau	Maple Leaf, No. 4	December 1, 1932
79.	William J. Hagney	Maple Leaf, No. 4	December 1, 1932
80.	Charles M. LeBlanc	Maple Leaf, No. 4	December 1, 1932
81.	Henry J. LeBlanc	Maple Leaf, No. 4	December 1, 1932
82.	Ezra J. Murray	Maple Leaf, No. 4	December 1, 1932
83.	William McCallum	Victoria, No. 4	February 17, 1935
84.	Walter Stewart	Victoria, No. 4	February 17, 1935
85.	Fidel Landry	Maple Leaf No. 4	March 22, 1938
86.	Roy Tipping	Victoria, No. 4	January 24, 1940
87.	James Burke	Victoria, No. 4	April 12, 1940
88.	Carl Linkletter	Standard	January 23, 1943
89.	David Muckle	Bayview	April 15, 1943
90.	Charles Gates	Bright Light	April 19, 1943
91.	James Porter	Standard	May 8, 1944
92.	Peter Gibbons	Bayview	February 8, 1950
93.	Frank B. Melanson	Cochrane	November 5, 1951
94.	Edward White	Bayview	March 12, 1957
95.	Frederick Burke	Bayview	November 3, 1957
96.	Havelock J Hoeg	Cochrane	January 6, 1959
97.	Elroy Dowe	Green Crow	May 3, 1960
98.	Hector G. McKeigan	Cochrane	March 29, 1978
99.	Hugh A. Stevens	Cochrane	August 17, 1978

TOTAL NUMBER OF DEATHS BY MINE (VILLAGE GIVEN IN PARENTHESIS):

1.	Victoria (River Hebert)	20	Bright Light	1
2.	Joggins	17		
3.	Scotia (Maccan)	9		
4.	Chignecto	9		
5.	Maple Leaf No. 4 (River Hebert)	8		
6.	Jubilee (Maccan Woods)	6		
7.	Marsh (River Hebert)	4		
8.	Cochrane (River Hebert)	4		
9.	Bayview (Joggins)	4		
10.	Minudie (River Hebert)	3	Kimberley	2
11.	St. George (Maccan)	3		
12.	Strathcona (River Hebert)	2		
13.	Black Diamond (Maccan)	2		
14.	Standard (River Hebert)	2		
15.	Great Northern (Maccan)	1		
16.	Lawson (Maccan)	1		
17.	Green Crow (Joggins)	1		
	TOTAL	99		

The monument also records the names of two men killed in the Westray mine disaster in Pictou County on May 9, 1992: Eric McIsaac and Danny Poplar.

APPENDIX D

POPULATION FIGURES

YEAR	ADVOCATE	JOGGINS	PORT GREVILLE/ FOX RIVER	RIVER HEBERT	PARRS BORO	SPRING HILL	SOUTH AMPTON
1991	196	491	181	773	1634	4373	244
1981	193	577	284	1090	1799	4896	
1971	219	777	286	1263	1807	5262	175
1961	311	909	356	1382	1834	5836	188
1951					1906	7138	
1941					1971	7170	
1931					1919	6355	
1921		1732			2161	5861	
1911		1648			2224	5713	
1901		1088			2705	5178	
1891					1909	4813	
1881					1206	900	

Appendix B

POPULATION FIGURES

Year	Advocate	Joggins	Port Greville	River Hebert/ Fox River	Parrs-borro	Spring-hill	South-ampton
1991	196	491	181	773	1634	4373	244
1981	193	577	284	1090	1799	4896	
1971	219	777	286	1263	1807	5262	175
1961	311	909	356	1382	1834	5836	188
1951				1906	7138		
1941				1971	7170		
1931				1919	6355		
1921		1732		2161	5861		
1911		1648		2224	5713		
1901		1088		2705	5178		
1891				1909	4813		
1881				1206	900		

Appendix C

SPRINGHILL MINE DEATHS

THE EXPLOSION OF 1956 AT SPRINGHILL MINE NO. 4 (39 DEATHS):

Joseph Crummey, Ben McLellan, Pleaman Pyke, Lester Nelson, David Vance, William Jones, Lester MacDonald, William Ferguson, Alex Spence, Alex Campbell, Logan Milton, Russell Morse, David Betts, Ernel Spence, George Ward, Ernest Boutilier, Dan Winters, Thomas Brown, Ralph Clarke, Gilbert Dakin, Victor Henwood, Leonard McCormick, William Tower, Avard Glennie, Harold Lewis, Donald Tabor, Kenneth Beaton, Clair Stiles, Vic Millard, Frank Allen, Floyd Beaton, Kenneth Clarke, Richard Ellis, Angus Hunter, Henry McLeod, Lester Fisher, Gerald Dawson, Burrell Pepperdine, and Ephraim Alderson.

THE BUMP OF 1958 AT SPRINGHILL MINE NO. 2 (75 DEATHS):

Fidele Allen, Ralph Aylward, Andrew Backa, Edward Bobbie, Bliss Bourgeois, Henry Brine, Percy Bryan, Charles Burton, George Canning, Cecil Cole, Hance Crowe, Harold Embree, Harry Embree, Harold Fraser, Joseph Gerhardt, Angus Gillis, Kenneth Goode, Harry Halliday, Cecil Harrison, Chesley Harrison, Harlan Henwood, Isaac Holloway, Hiram Hunter, Wylie Hunter, Warren Hyatt, John Jackson, William Jewkes, Abbey LeBlanc, Alfred Legere, Gilbert Livingstone, Arthur MacDonald, Edward MacDonald, Harold MacDonald, Roy MacFarlane, Frank MacKenzie, Edwin MacKinnon, Charles MacLeod, Clarence MacLeod, Edward MacLeod, Frank MacLeod, Robert MacLeod, Varley MacLeod, Harold McNutt, John Maddison, Thomas Marshall, Bernard Miller, Carl Mooring, Fred Nicholson, Henry O'Brien, Robert Perrin, Stirling Porter, Harold Raper, Percy Rector, Joseph Reid, Layton Reid, Lester Reid, Wesley Reynolds, Ernest Rolfe, Charles Ross, Philip Ross, Robert Ross, St. Clair Ross, William Smith, Percy Spence, Eldon Stevens, William Stevenson, Hollis Tabor, Monty Tabor, Raymond Tabor, Henry Teed, William Turnbull, George Welch, Albert White, Carl White.

Acknowledgements

I would like to thank all those who were so helpful; without them this book would not have been possible—thanks again.

Sharon Gould, Upper Nappan; Irene Melanson, West Amherst; Sandra Hunter, Leamington; Evelyn Brown, Oxford; Debbie Boudreau, Lower Cove; Marion Kyle, Lakelands, (a great help); Mary Willa Littler, Springhill; Willo Thompson, Southampton; Helen Sims, H.R.S. Computer Services, Joggins; Karen Dickinson, West Brook; Judy Jollymore, River Hebert; Helen Winters, Oxford; Gussie Morris, Advocate Harbour; Muriel Fletcher, Port Greville; Michelle Lelievre, Pictou County; Jannet (Skinner) Wilson, Moncton; Myrtle Chappell, Fenwick; Marion Kindervater, Upper Musquodoboit; Doris Gilroy, Mapleton; Dayle Hoeg, South Athol; Deloras MacFarlane, Parrsboro; Helen (Mrs. Lovitt) Carter, Point de Bute; Jeanne Fulton, Port Greville; Kerry Lawrence, Southampton; Jim Lawrence, Calgary, Alberta; Ralph Thompson, Amherst; George Blenkhorn, Southampton; Lawson Brown, Southampton; Gary Porter, Strathcona; Reg (Bud) Johnston, River Hebert, (a great help); Leo Burke Sr., Ontario; Norman and Gordon Fullerton, Southampton; William McLellan, River Hebert; Troy White, Amherst; Walter Newcombe, Brookdale; Earl Pettigrew, Tidnish; Conrad Byers, Parrsboro; Eldon George, Parrsboro; Johnson McPhee, Parrsboro; Russell Fisher, Springhill; Brock Kindervater, Hinton, Alberta; Cecil Brown, Mapleton; Norman Harrison, Leamington (a great help); Morley Greer, River Hebert; Kim Greer, River Hebert; Eric Long, Valley; John Munro, Ragged Reef and Amherst; Peter Cottingham, Polka Dot Antiques, Amherst; Gary Murphy, Ragged Reef; John Reid, Joggins; Loring Kerr, Port Greville; Mr. and Mrs. Roger Harrison, Minudie; Kirk and Beth Reid, Advocate Harbour; Charles and Lois Davison, Newville Road; David and Janette Dinaut, Port Greville; John and Barb Reid, Maccan (a great help); Stephen and Sharon Fraser, Advocate Harbour; Robert and Helen Gilroy, Leamington; William and Beryl Atkinson, Amherst; Roy and Marjorie Hoeg, South Athol.

the writings of:

M. Hunter, 1901
Annie Thompson, 1966, West Brook
Millie Smith, 1970, Southampton
Mrs. Donald Harrison
John G. McKay, Amherst
Ernest E. Coates, Fenwick
Laura Goodwin, Southampton

Harry Burke, Joggins
Ella Fraser, Spencers Island
Paige Mahar
Mrs. Edwin Davison, Newville
Conrad Byers, Parrsboro
Stanley Spicer, Spencers Island
David Beatty, Sackville, N.B.

Many thanks are extended to the staff at:

The Amherst Daily News & Citizen
Heritage Models, River Hebert
River Hebert Miners Museum
Cumberland County Museum
Acadian Printers
King Seaman Museum
Mt. Allison University
Joggins Fossil Centre
Age of Sail Museum, Port Greville
Museum of Industry, Pictou County
Amherst Library, Four Fathers
Hector Centre, Pictou County

Image Sources

The Amherst Daily News, The Halifax Chronicle-Herald, Parrsboro Record, Springhill Record: pages 103, 133, 135, 136, 137
Anne Murray Centre: page 138
Brown, Roger: pages i, viii, 7, 12(lower), 14, 26, 42, 47(upper), 50(lower), 58, 59, 60, 62, 63, 64, 66(lower), 67, 70(lower), 74(lower), 85, 86(lower), 89, 90, 91(lower), 92, 93(upper), 96, 97(lower), 104, 118, 127, 129, 130, 143(lower), 145
Byers, Conrad: pages 10(lower), 16(upper), 32, 34, 84, 88(lower), 95, 108, 120, 122, 123(lower), 139(upper)
Carter, Mrs. Lovitt: page 27
Cottingham, Peter (Polka Dot Antiques, Amherst): page 125(lower)
Davison, Charles: pages 53, 74(upper), 132
Dinaut, David and Janette: page 119(lower)
Fisher, Russell: page 142(lower)
Fletcher, Muriel: page 121(upper)
Fraser, Stephen and Sharon: page 117
Fullerton, Gordon and Norman: pages 91(upper), 146(upper)
Fulton, Jeanne: pages 28, 73, 94(lower), 112, 114, 115, 121(lower)
George, Eldon (Parrsboro Rock & Mineral Shop): pages 87, 88(upper), 93(lower)
Gilroy, Doris: page 70(upper)
Gilroy, Robert and Helen: pages 105, 144, 146(lower)
Greer, Kim: pages 18(upper), 126
Greer, Morley: pages 25(lower), 49(lower), 123(upper)
Hoeg, Dayle: pages 35, 36
Hoeg, Roy and Marjorie: pages 66(upper), 68, 75
Hunter, Sandra: page 65
Johnston, Bud (Heritage Models, River Hebert): pages 4, 11(lower), 12(upper), 13(lower), 15, 16(lower), 17, 20, 22, 24
Kerr, Loring: page 119,
King Seaman Museum: pages 45, 46
Kyle, Marion: pages 81, 82, 83, 86(upper), 116(upper), 139(lower), 140, 141, 142(upper), 143(upper)
MacFarlane, Deloras: page 113
McKay, John G.: page 31
McPhee, Johnson, pages 94(upper), 97(upper), 98, 99, 100, 101, 102
Melanson, Irene: page 37
Morris, Augusta: pages 109, 110, 111(upper), 116(lower), 124
Munroe, John: page 49(upper)
Murphy, Gary: pages 19(lower), 21(lower)
New Brunswick Museum: page 78
Newcombe, Walter (Station Street Flea Market, Amherst): page 111(lower)
Reid, John and Barb: pages 8, 9, 10, 30
Reid, John (Joggins): pages 19(upper), 29
River Hebert Miners' Museum: pages 13(upper), 23, 39(upper), 48(lower)
Saint Denis Roman Catholic Church (Minudie): page 5
Sims, Helen (H.R.S. Computer Services of Joggins): pages 1, 11(upper), 18(lower), 21(upper), 25(upper), 39(lower), 50(upper)